ENVIRONMENTAL CHALLENGES AND SUSTAINABLE FUTURE

ENVIRONMENTAL CHALLENGES AND SUSTAINABLE FUTURE

Gopal Bhargava

Publications

KALPAZ PUBLICATIONS

DELHI–110052

ENVIRONMENTAL CHALLENGES AND SUSTAINABLE FUTURE
(Environment)

Rs. 740

ISBN: 81-7835-211-7

Published in 2003 in India by
Kalpaz Publications
C-30, Satyawati Nagar
Delhi-110052
E-mail: kalpaz@hotmail.com

Laser Typesetting at: Varsha Graphics, Delhi
Printed at: Mehra Offset Press, Delhi

Contents

Preface

The World Summit on Sustainable Development, held in Rio de Janeiro in 1992, resulted in a world commitment to managing the environment within the process of development – not as an afterthought. At the core of the 1992 WDR, was a recognition that the environment, or a country's natural capital, can provide a sustainable flow of good and services that are needed for development. The main environmental challenges that the WDR identified were water and sanitation, energy and industrial growth, rural development and international environmental issues. The WDR made the case that the links between environment and development were based fundamentally on the use and management of natural resources and that the creation of environmental externalities was a hindrance in the process of economic development.

Most developing countries have made progress in establishing a legal and regulatory framework for environmental protection, over the past two decades. They have set up environmental protection agencies, promulgated standards, and put in place at least rudimentary monitoring and enforcement systems. However, environmental quality has deteriorated for most indicators and in most of the region. Many observers of the worsening situation pronounce regulation unworkable, whether it be energy intensity standards, appliance standards, emissions or effluent discharge standards, treatment standards, or mandatory technology adoption.

Environmental management are now a regular feature of political discourse throughout the developing world. Over the past fifteen years in particular there has been a surge of public interest and action in this area. Today the level of citizen participation in environmental groups in the developing world compares. The rise of civic environmentalism has taken place in a broader context of increasing civic organization throughout the developing world. Non-environmental organizations include professional, religious, artistic, charity, and sports associations, as well as political parties.

While recognizing that the false dichotomy of environment and development still exists, it is equally important to recognize that developing countries today have more "role models" amongst other developing countries where interventions have paid off. Communicating these success effectively will be critical to building institutions and attitudes that can stand up to political pressures, address equity concerns and move this dialogue more aggressively towards economic policies that are not only theoretically sound, but can indeed move closer to identifying sustainable development outcomes.

Editor

1

Environment and Civil Society

Some sustainable issues development like government policies, market incentives, and public pressure need to align to push reluctant public and private sector enterprises to waste less, pollute less, and meet more of the needs of low-income households, women, and socially excluded groups are in the developing stage. However, sustainable development will not result solely from a fortuitous conjunction of correct pricing, sound regulation and enforcement, and inevitably sporadic public pressure.

The challenge is too great; the gap between crisis and potential achievements of environmental policy shifts is widening daily. Only fundamental shifts in culture, in the mindsets of enterprise managers, government functionaries, and owners of capital, can extricate Asia and the Pacific from a morass of obsolete technologies and process designs and overcentralized infrastructural behemoths, and put it on the path to a revolutionary transformation of production, consumption, and distribution of resources.

Prices of resources such as energy, water, minerals, and timber should reflect marginal costs of their provision at point of purchase plus the external costs imposed on society by resource extraction, transport, conversion, and use. Consumer product prices should fully reflect the sum of the externalities caused by production, transport, and consumption.

However, while economists can argue for decades over how to get prices right, no country in the world has made more than tentative and modest steps toward full price reform. Prices of energy, water, and almost all products and services (including transport) would have to significantly increase, and few countries can even achieve marginal-cost energy and opportunity-cost water pricing, let alone internalize detrimental social and environmental impacts. General, non-targeted subsidies on all fuels, electricity, fertilizer, water, and pesticides should first be removed. Although they have been to some extent, attempts to implement the polluter-pays principle have been generally ineffective because pollution charges are too low. Pollution charges are generally meant to raise revenue rather than to spur reduction in pollution.

One major difficulty in raising energy prices to a level that reflects the social and environmental externalities caused by energy use is populist politicians' belief or claim that the high prices hurt the people. Thus, many countries in the region subsidize diesel and kerosene. As a consequence, diesel is a major source of health-threatening air pollution in Asian cities.

The benefits of diesel subsidies, however, flow mainly to the nonpoor. There are far more effective ways of achieving real income gains for the poor without threatening their lives and health in the process. Energy taxes would raise revenue while both stimulating energy efficiency and reducing environmental and social externalities.

In Asia and the Pacific, such taxes would logically focus on the externalities of most immediate and direct concern in the region: sulfur, heavy metals, particulates, and site dislocations. The transport infrastructure, roads are highly subsidized and most countries lack pricing policies that attempt to incorporate environmental and social impacts in road-user charges. Ensuring the proper structure and level of such charges is a complex process, but light vehicles assume all charges as incorporated in a fuel tax. The objective is to move toward more efficient use of available transport capacity, choice of vehicles and fuels, split of traffic among modes, and maintenance of infrastructure.

Road traffic involves three external effects: congestion, mainly in urban areas, environmental externalities, which are functions of amount and type of fuel, engine type, and noise-producing characteristics of vehicle and engine; and road damage, caused primarily by heavy vehicles and not captured by a fuel tax. Fuel taxes are a reasonable way to capture environmental externalities related to air pollution. Localized air pollution externalities in developing countries may add a further $0.25/l, not including the cost of accidents, global environmental impacts, and oil-related military expenditures.

In OECD countries- which are plagued by technological inertia and by oil and car cartels-that high prices do not magically transform transport systems. State intervention is thus necessary to promote national land-use planning, public transport, and alternative transport investments such as walkways, bicycle systems, and alternative fuel vehicles and fuelling distribution systems. At stake are not fuzzy aesthetic or ecological values, but the idea that economics must be viewed from the perspective of society rather than of energy cartels. A 20 percent reduction in suspended particulate matter in Bangkok alone would produce annual health benefits in lower morbidity and mortality estimated at over $1 billion.

Charges reflecting the cost of fuel use to human health alone would be significant. They would not only represent a major revenue source but also effect a progressive income impact if used, for example, to subsidize mass transport, construction of walkways and bicycle lanes connecting low-income settlements to transport nodes, the retraining of coal miners, research and development in energy efficiency and clean production, promotion of energy efficiency investments, and the switch from two-stroke to four-stroke motorcycles.

Prices can be complemented by policies to make markets more competitive by breaking up cartels in strategic sectors such as energy, promoting entry of small and medium-sized enterprises into energy service businesses, and creating markets in "bads". The aim of getting prices right should be to reduce waste and pollution and in an economically efficient manner while promoting long-term structural transformation toward zero waste. Effluent and emission charges and tradable emissions or effluent permits can, theoretically, help increase efficiency and raise revenue.

Charges allow firms the flexibility to make decisions based on the actual costs of abatement to the firm and should lead to lower aggregate costs of compliance than use of regulatory standards based on effluent or technology criteria. Market-based instruments also offer the theoretical possibility for dynamic efficiency gains as they provide a continuous incentive for environmental improvement, including process changes. However, it is difficult to rely on market-based instruments, which explains their rarity in the region.

In Singapore, road and vehicle taxes that has implemented in an effort to prevent congestion problems such as those affecting large urban areas in neighbouring countries. In addition to import duties of 45 percent and registration fees of S$1,000 ($710), Singapore imposes an Additional Registration Fee (ARF) based on the market value of the vehicle. The ARF rose from 15 percent in 1968 to 175 percent in 1983, before falling to 150 percent in 1991.

The annual road tax is based on the engine capacity of the vehicle. These rates have risen significantly since the early 1970s. The ARF is reduced when an old vehicle is scrapped and a new one purchased, to discourage ownership of older, high-emitting vehicles and to limit the used-car market. In 1990, Singapore implemented a quota scheme under which vehicle owners are required to have Certificates of Entitlement (COEs), which are valid for 10 years and obtained at monthly public auctions. Owners of vehicles more than 10 years old are required to pay the prevailing quota price.

The COE requirement enables the Government to determine the total number of vehicles in circulation based on the country's road capacity. COEs cannot be transferred. COE prices have increased rapidly. For cars with a capacity over 2,000 cc, they have risen from S$528 ($375) when they were introduced in 1990 to S$17,600 ($12,500) in 1992 to over S$100,000 ($70,000) in 1994. The off-peak car scheme is designed to reduce congestion. Off-peak car buyers enjoy tax rebates on the registration fee, import duty, and COE premium, up to a maximum of S$17,000 ($12,000). They are also entitled to S$570 ($300) reductions in road tax.

Off-peak cars can be used on Sundays and public holidays and during off-peak hours. Off-peak vehicles are clearly marked by red number plates that are welded onto the vehicle and sealed by an authorized inspection center. To drive the vehicle outside the authorized times, a S$20 ($14) day license has to be displayed on the windshield. Owners get five free day licenses a year. The unleaded gasoline tax is S$0.60 ($0.43) per liter or 50 percent of pump prices (including taxes), whichever is higher. One problem that arose as a result of the taxes was that motorists purchased fuel in neighbouring Malaysia, where a liter of gasoline was about S$0.50 ($0.35) cheaper.

Singapore countered this practice by requiring all vehicles leaving the country to have their gasoline tanks at least ¾ full. As the main operator of parking facilities, the Government also imposes relatively high parking fees. Parking charges within the Central Business District (CBD) are S$0.90 ($0.64) per half hour during office hours. Outside the CBD, charges are S$0.45 ($0.32) per half hour.

In 1975, the Area Licensing Scheme (ALS) was adopted and modified in 1989 and 1994 to reduce congestion in the CBD. Cars entering the CBD need a license: a part-day license is S$2 ($1.40) for entry into the CBD during off-peak hours and a whole-day license is S$3 for use between 7:15 a.m. and 6:30 p.m.). The ALS has had a large impact on peak-hour traffic, resulting by end-1975 in a 71 percent decrease in the number of private vehicles entering the restricted zone between 7:30 and 10:15 a.m. Public transportation became the preferred mode of transportation after the introduction of the ALS.

The 1989 expansion of the system to evening peak hours resulted in further traffic decreases and increases in average speeds of 10.8 percent in morning peak hours and 30.4 percent during the evening peak period. Road pricing is electronically handled, with tolls deducted from vehicle transponders and authorities notified of toll violators. The COE and other measures are credited with significantly limiting the number of vehicles in Singapore. It has been estimated that without vehicle ownership and use disincentives, the number of vehicles in Singapore would have been 400,000 by 1992 instead of the actual number of 274,000.

The road-pricing program, combined with other charges on vehicle ownership, has dramatically reduced traffic congestion and air pollution in downtown Singapore. Business activities and rents downtown have not suffered. Vehicle taxes accounted for 23 percent of total government tax revenue in 1994.

Industrial Pollution

Most developing countries have made progress in establishing a legal and regulatory framework for environmental protection, over the past two decades. They have set up environmental protection agencies, promulgated standards and regulations, and put in place at least rudimentary monitoring and enforcement systems. As is all too well known, however, environmental quality has deteriorated for most indicators and in most of the region. Many observers of the worsening situation pronounce regulation unworkable, whether it be energy intensity standards, appliance standards, emissions or effluent discharge standards, treatment standards, or mandatory technology adoption.

Ironically, every attempt at regulation has produced success stories in different countries. Unfortunately, the successes are dwarfed by the magnitude of the task.

In the early 1960s the state of Gujarat, India, is made up of about 250 small-scale industries. Many of the industries are chemical producers. Nandavari estate has a long history of water pollution problems. The pollution-related damage has had an impact on both drinking and irrigation water as well as other livelihood uses such as local fishing. There are records of various complaints lodged by the local people about the degraded quality of local water by industrial effluents. Consumer groups have filed complaints in the local court.

Over the years, the State Pollution Control Board has filed about 200 cases against the various factories. Two prior attempts at solutions, both initiatives of the public sector-a state-sponsored common treatment plant (CTP) and an effluent channel project-failed to solve the problem as industrial influents to the CTP were such that it could not function adequately. Complaints continued. However, no further remedial action was taken by the Government until a local environmentalist lawyer filed a public interest litigation case against the factories. As a result, the court ordered the closure of the factories in 1995.

The court order encouraged the Nandesari Industries Association to take steps to avoid closure while also seeking a cost-effective solution. The result was a collective action, which proved to be a cost-effective alternative institutional arrangement combining both (i) industry-specific primary treatment for those industries unable to meet CTP influent requirements, and (ii) additional treatment by an improved CTP. The CTP was taken over by the industrial association, which upgraded and now operates the treatment plant.

Investigations suggest that the industries cooperate because the collective action allows them to capture the economies of scale a CTP affords and because they have mutual expectations regarding their behaviour. The latter is facilitated by the institutional arrangements put in place by the association, ensuring a fair sharing of costs, which are based on both quantity and quality of the effluents treated at the CTP.

In Korea, domestically owned factories do better than foreign-owned ones, but in some other countries the reverse is true. Some studies find that State-owned enterprises are more subject to and responsive to regulatory pressure than private and collective enterprises. Most studies from Asia show that regulation matters and that response to regulation is surprisingly uneven.

A significant determinant of firms' responsiveness to regulation is community pressure, which also appears to be a major determinant of regulatory pressure and application of pollution devices where, in spite of uniform national devices, local devices are tailored to the intensity of local community pressure. Thus, application of the pollution levy system is neither arbitrary nor ineffective as sometimes claimed, but represents local adaptation to local conditions.

Studies shows that Asian countries local income and education levels are predictors of the effectiveness of regulation at the local level. A national regulatory system is necessary and that its local application is roughly a function of pollution damage and community education, and bargaining power. The effective environmental improvement regulations can have an impact in the hands of an informed and aroused citizenry, as in oecd countries.

Unfortunately, the implication is that rich and politically powerful communities can protect themselves while low-income neighbourhoods gasp, wheeze, and vomit in the filth and dregs of the industrial revolution. Besides community pressure and regulatory enforcement, the third key element in industrial environment management is the enterprise itself.

Asia and the Pacific producing, monitoring, and enforcing standards and regulations, to providing information and ceding power to the public. Environmental policies can broaden their scope to consider firms as organizations subject to the power of information, social pressure, economic forces, and regulatory actions.

Public disclosure of factories' environmental performance, voluntary participation of firms in environmental ranking schemes, pollution discharge registries open to public view, community monitoring of environmental quality, and environmentally focused community organizing could be elements of an effective approach that broadens the range of action away from sole reliance on market and State. Overreliance on such a strategy, however, is subject to severe limitations. As already mentioned, citizen involvement has the most impact in higher-income and better-educated areas. The phenomenon not only suggests that low-income communities are at a disadvantage, but also that increasing the environmental voice of richer communities will, over time, force even more relocation of the dirtiest, most disruptive activities to low-income areas.

For example, the announcement by plantation owners in the Philippines that they would reduce pesticide use would have been welcome, but it avoids the real issue of sustainable agriculture-plantation monocropping is unsustainable and overapplication of pesticides is only one part of the problem as long as high-impact monocropping continues.

It is true that it is necessary to get prices right, foster market competition, and create a credible regulatory apparatus. At best, however, these measures are initial steps toward sustainable industrial development that will merely achieve incremental gains within firms and further reallocate social and environmental externalities toward low-income and marginal communities. Even the removal of barriers, however, is no easy task. Sustainable development, environmental quality, and equity are political and therefore cultural issues. So, while prices are adjusted to reflect full resource costs and externalities, and markets are regulated or deregulated depending on the circumstances, agents of change must be active on both the political and cultural fronts.

Scientists, technology developers, and management must recognize their value-laden contexts and shift from reductionist, linear, incremental thinking to a holistic, cyclical, revolutionary outlook. Rights of low-income communities and socially excluded groups must be recognized and upheld. They include (i) rights to life, health, and safety; (ii) the right of access to resources, including land; and (iii) the right to a political voice.

Effective legal systems and governments are imperative for equity and environmental protection. Neoliberal arguments that have led to the weakening of government's capability to cope with poverty reduction and environmental protection are misguided and have damaged society and nature.

The market has a supportive role to play but only under a strong, democratic system where political decisions are made through a transparent political process rather than through an "impersonal" market characterized by egregious concentration of monopoly power and widespread market imperfections, asymmetries in access to capital and information, and refusal to account for social and environmental externalities. Political decisions made transparently need to be followed by accountable implementation in a process where citizen beneficiaries can trust State intentions; depend upon the promises of politicians and bureaucrats; have recognized rights to life, health, and livelihood; and have recourse when their rights are violated.

Freedom of the poor, of women, of socially excluded groups to associate and organize is essential to governmental accountability. The recognition of and support to many loci of civic action is a necessary complement to an honest and effective State and a competitive and regulated market. Social mobilization must achieve numerous purposes:

(i) reduce individual vulnerability;

(ii) increase incomes;

(iii) build social cohesiveness;

(iv) protect the environment;

(v) resolve conflict;

(vi) effectively and efficiently deliver public services; and

(vii) give voice and creative expression to the silent.

Some developers can get rough when poor, informal settlements stand between them and incredible profits from land grabbing. Illegal loggers have strong incentives to be violent with tree huggers. The challenge is for governments to align themselves with expressed local desire and to reflect broad-based social values rather than narrow greed-based interests. Transparency and accountability of government and corporate behaviour are dependent upon continuous and vigorous political pressure from below.

Donor-advanced decentralization has been a guise for reducing government budgets and responsibilities, avoiding regulation of illegal practices, shifting risks to localities and individuals, undermining government line agencies, and enhancing the power of local elites. The weakening of the power and authority of government institutions fits in perfectly well with the needs of polluting industries, ravaging loggers, foreign fishing fleets, and global investors seeking to penetrate new markets. The challenge is for the rough and tumble of democracy to force accountability and transparency.

This challenge must be not only devolution of authority to governments closer to the people affected by decisions, but also enforced legal rights for individuals and groups to participate in political and economic processes. This implies a far greater degree of social inclusivity than is common in many countries today, an inclusivity that threatens social norms justifying discrimination, inequity, abject poverty, and violence against the weak and powerless. Equity of legal rights dovetails with the citizenry's vigilance necessary to keep governments, corporations, and NGOs accountable to those they supposedly serve.

New local government legislation in Bangladesh provides for direct election to one third reserved seats for women in all four tiers of local government. There are now more than 12,000 women directly elected as Union Council members. The challenge now is to ensure these women's full and effective participation in local governance. The 73rd Constitutional Amendment in India mandates participation of women in local government, and the results, although mixed, are encouraging to the view that participation of the socially excluded in politics leads to empowerment.

Of equal or greater importance to these attempts to broaden access to government are the many loci of organizing that may make "Voting to Abstain," as a recent cover story in India Today put it, a more powerful force for social transformation than incorporation of the demands of the excluded into election campaigning. Both venues of struggle are important and necessary to good governance.

The transformation of "hopeless pits of aridity, despair, alcoholism, and out-migration" into oases of prosperity such as has happened in Ralegaon Siddhi, Maharashtra, India, and so many other self-help villages of Asia and the Pacific offers fertile ground for optimism.

How much faster the process could occur with State support and the participation of NGOs that act as true partners of community organizations. Central and local governments, NGOs, and community organizations of many stripes all have roles to play in complex, nuanced, interactive, and dynamic ways that challenge simplistic characterizations about decentralization, participation, and civil society.

Self-reliance, self-confidence, and self-knowledge are all keys to social transformation and to new ways of thinking and doing-ways that give value to all living beings, credit to everyone's opinions, and delight in the beauty of an intact ecosystem and vital social system. Chaos and conflict may be part of the process, but as the creative energies of all are freed from the oppression of hierarchy and exploitation, societies may emerge that nurture life and trust in the wisdom of the ancients.

In market, consumers and producers need to know what is available, what are the lifetime costs, and what will be the expected health, safety, social, and environmental implications of their choices. Information is the key to the functioning of effective markets. Lack of information hides corruption, collusion, anti-competitive practices, private profiteering from exploitation of public resources, and negative social and environmental impacts.

The information age will contribute to sustainable development if information is not monopolized or distorted. State intervention is required to encourage openness and accountability of all market players, whether public or private. New rules of the industrial game include the following:

— Industries must get a license to operate from the stakeholders-the workforce, neighbours, inhabitants of the ecosystem they affect, and representatives of future generations. The licensing process is essentially a public sharing of information on costs, benefits, and risks.

— Corporate governance and accountability can be enhanced by community access to information, especially on health, safety, and the environment.

— National policies can only be feasible and effective if an international compact is set within a global framework that includes both micro-level investor responsibilities and macro-level sustainable development objectives. Minimum health, safety, and environmental standards need to be agreed upon. Such standards allow higher standards as national or sub-national options. The process of developing standards must be open and inclusive, and any standards must include information disclosure responsibilities. Any global framework should cover both domestic and foreign investments.

A major challenge for sustainable development in the developing countries of Asia and the Pacific is thinking through the alternatives. The industrialized countries achieve hegemony not merely through military and economic might but also through intellectual and cultural leadership. Such leadership inevitably puts their interests first. It is futile to complain about it. The challenge is to build capacity within the region to seek and construct sustainable development strategies most appropriate for the diverse conditions prevailing in the region. Sustainable development think tanks would take up the following issues:

— women and gender issues;
— social exclusion;
— management, norms, values, structures;
— conflict resolution;
— organizing the powerless;
— links to grassroots groups such as people's organizations, advocacy organizations, and grassroots research groups;
— values, ethics, contemplation, spirituality;
— legal aid, public interest litigation;
— Asian Coalition for Housing Rights low-income group exchanges;
— international links to support a community orientation;
— children;
— teaching environmental and human rights law; and
— achieving accountability and transparency in government.

Poverty reduction has been proclaimed as the overarching goal or a high-priority objective of many international financial institutions, bilateral agencies, and national governments. However, the key question is how to achieve this noble goal. For example, ADB identifies the following as pillars of its poverty-reduction strategy: (i) pro-poor economic growth; (ii) good governance; and (iii) social development (including empowerment of women and collaboration with NGOs and civil society). Countries and their development partners have to devote a great deal of time and attention to strengthen these three pillars to reduce poverty in Asia. They should put greater emphasis on (i) agriculture and rural development, (ii) environment and natural resources management, and (iii) transport and energy development to meet the basic needs of the poor.

The implementation of Tobin taxation will generate a minimum of $150 billion/annually in additional resources for poverty reduction. In March 1999, the Canadian Parliament approved the concept of the Tobin tax. There is also a global initiative to get the idea approved by all donors. Thereafter, implementation and monitoring may be done by international financing institutions or through some other suitable arrangement. Since 80 percent of all the global currency trading takes place in seven financial centres in the world, implementation and monitoring should not be so difficult.

In 1994, the African governments called for a one-third reduction on military spending to release funds for public spending on social programs. Similar appeal has been made in many other global and regional forums. UNDP has called the use of scarce resources for military expenditure "the most shocking example of states' use of power contrary to the interests of poor people. These three international initiatives must be associated with three other regional actions of (i) pro-poor economic growth, (ii) social development, and (iii) good governance to bring Asia out of abject poverty by 2015.

Information on water conditions is generally lacking, however, and rates of natural resource loss often exceed local government's attempts to remedy environmental problems. To remedy this situation and to build community awareness of water quality problems, community groups attended workshops where they learned how to monitor water quality and watershed stability in streams and lakes using inexpensive, simple test kits, and other basic and portable analytical tools.

The thousands of water quality samples collected by the citizens revealed a clear west-to-east gradient of degradation across four subwatersheds of the municipality. Citizens' active involvement in finding solutions for conserving local water supplies exceed government's capacity to measure conditions, identify specific problems, and decide upon a proper course of action. Especially in rapidly and irreversibly deteriorating watersheds, application of partly understood conservation practices, with full community involvement, may work better than waiting for a "complete" scientific analysis.

Many developing Asian economies, like Taipei, China, has a large segment of industrial production that comes from small and medium-sized enterprises. Among approximately 90,000 firms in Taipei, China, more than 96 percent are small and medium-sized enterprise- those with a capital investment of less than $1.5 million, total assets less than $4.5 million, or employees less than 200. These firms together generate about 50 percent of the gross production of all business enterprises in Taipei, China. Although small on a per-facility basis, the gross amount of waste from small and medium-sized enterprises is substantial. Moreover, as many of the enterprises are scattered throughout commercial and residential areas, their environmental and health impact on the public is especially severe.

Small and medium-sized enterprises generally have little financial, technical, or man-power capability to implement adequate environmental measures. The public exerts substantially less pressure on them than on large firms in environmental, health, and safety issues. As a consequence, small and medium-sized enterprises are comparatively less active in environmental programs.

The corporate synergy system is a management approach that involves forming partnerships among business organizations to achieve specific goals. Usually established within supply chains, corporate synergy systems are initiated under the leadership of large companies, where the upstream suppliers and downstream buyers in the chains are organized to work together to achieve common goals-in this case, cleaner production.

Under the leadership of large companies, supply chains are becoming more integrated in terms of decision making and planning, with greater exchange of information among chain members. In addition to cost containment and quality assurance, many large firms have begun to work with the upstream suppliers and downstream buyers to reduce their environmental responsibilities. Central firms of a corporate synergy system reward "good" suppliers by providing special credit treatment, free staff training, and/or relaxed quality audit requirements. Eventually, through the operation of a corporate synergy system, large companies use a variety of parameters, including product quality, financial strength, and environmental performance, to rank their suppliers.

In 1995, two corporate synergy systems have been established in Taipei, China, to promote clean production in supply chains. TECO Electric and Machinery Co. Ltd., one of the largest electrical appliance and equipment manufacturers in the country, initiated the first system. Before corporate synergy system clean production started, many of TECO's plants had implemented clean production programs and realized their financial and environmental benefits.

TECO's top management was convinced that further substantial gains could only be realized if its suppliers adopted clean production measures through corporate synergy system clean production. At the first stage of the system's organization, however, many of TECO's suppliers were either not interested or only passively participating in the program. In the TECO corporate synergy system clean production system, two of TECO's electric appliance plants and two electric motor assembly plants served as the central firms. There are 12 satellite firms representing suppliers of data-processing equipment, printed circuit boards, parts molding, etc.

Established in July 1996 with Cheng-Loong's Tayuan Paper Mill and Hsinchu Paper Mill as central firms, the corporate synergy system consists of 10 upstream suppliers that provide waste paper, machinery, chemical, energy, and transportation services, and 3 downstream buyers who are paper container manufacturers. With small and medium-sized enterprises making up more than 90 percent of participating firms, the Cheng-Loong system implemented 868 clean-production options from July 1996 to June 1997. The participating firms invested $991,000 in clean production measures, resulting in an annual benefit of $3.5 million.

In India, Surat has over 250 polyster-dyeing and -printing houses, which employ equipment, dyestuffs, and auxiliaries that are highly comparable to those found in

Europe. The supply industry for chemicals and dyestuffs and a few entrepreneurs have taken the lead in organizing the Waste Minimization Group (WMG), which brings together textile processors and their suppliers. WMG aims to create awareness of environmental problems caused by textile wet processing, sharing WMG experiences and fostering WMG-driven innovations. An early success is the development of the extremely low liquor jet dyeing machine for low-grammage cloth. The machine is approximately three times as energy and water efficient as comparable equipment made in Europe.

Need for Environmental Protection

Indian Constitution emphasizes the need for environmental protection. Article 48-A states: "The State shall endeavour to protect and improve the environment and to safeguard the forests and wildlife of the Country." Article 51-A(g) confers a fundamental duty upon every citizen of India to protect the environment: "It shall be the duty of every citizen of India to protect and improve the natural environment, including forests, lakes, rivers and wild life, and to have compassion for living creatures."

Judiciary of India has played a major role in protecting the environment and natural resources and in imparting environmental justice to the victims of pollution. The Supreme Court of India has broadened the scope and ambit of the right to life and liberty to include important issues of public interest, including the right to a healthy environment. The Indian Supreme Court has interpreted fundamental rights as enshrined in the Indian

Constitution as a repository of human rights and the source of law for a just society. Thus, the right to life has been interpreted as the right to a decent living, the right to minimum wages, and the right to a healthy environment. Through landmark judgments, the Supreme Court has given recognition to the polluter-pays, precautionary, and public trust principles.

The liberalization of the rule of locus standi has enabled environmentally conscious, public-spirited individuals and groups an access even to the highest court of the land. Public interest litigation may be viewed as a collaborative effort on the part of the petitioner to secure observance of constitutional and legal rights conferred upon vulnerable sections of the community and to bring social justice to them. So far, more than 5,000 cases of prosecution against polluters are pending in different courts in India.

A number of writ petitions filed by individuals and voluntary organizations are also pending in the High Courts and the Supreme Court of India.

For example, air pollution and acid rain from an oil refinery and other industries in Agra are damaging the Taj Mahal's surface. Based on public interest litigation seeking the closure or shifting of pollution sources, the court ruled that the burden of proof was on the industry to show it was environmentally benign. It ruled that 292 coke-based industries should shift to natural gas or face closure, directed the stoppage of coal supplies to these industries, and recommended moving small-scale polluting industries and providing natural gas to industries and the oil refinery in the region.

Forests in India, are damaged by several environmental pollution. Agriculture farms have been established in 85,000 ha of ecologically fragile land along the coastline. Despite environmental law and landmark judgments delivered by the Supreme Court and some of the High Courts of India, the situation has not changed dramatically on the environmental front. The implementation of the courts' orders rests with the executive, which remains indifferent to its responsibility to enforce environmental regulations.

In Thailand, in the year 1964, the province had 87 percent forest or over 1 million ha of forest. Improved transportation and the rapid development of Thailand brought broad social and economic changes to Nan. Natural resources were rapidly depleted and, by 1993, only 480,000 ha of forest cover were left. Most of the deforestation was due to commercial logging. After an area had been logged and burned, it was planted to export crops such as corn, mung, cassava, cotton, tobacco, and sugarcane. The entire economy became export oriented.

Deforestation had serious effects on watersheds, resulting in prolonged dry seasons, and erosion and flooding during the wet season. Watershed areas, timber reserves, and biodiversity were destroyed, and environmental conditions were worsened by agricultural chemical use. Farmers fell deeper into debt through adoption of high-input export-oriented agriculture.

Today they feel more vulnerable than ever to shock. Rapid social and economic changes, and concern for preserving the environment and local culture have led to the creation of Hak Muang Nan (HMN), a network of development workers, monks, community leaders, and government officials.

HMN has a strong local focus and has led protest actions against logging companies, developed programs and provided credit to promote sustainable agriculture, and provided support services. By the late 1990s, HMN had grown from 14 community organizations to 170, with 39 community forests and 53 river conservation zones. HMN is also strongly oriented toward livelihood projects, particularly credit unions, vocational training, nonfarm employment, and sustainable agriculture promotion. It is active in promoting diversified farming, including orchards, ponds, small-animal raising, and community forestry.

In India, Unionization of Women Tobacco Workers in Kheda, Gujarat, India, Kheda is one of the richest districts in Gujarat, but it is also one where inequality is severe. At the lower end are improverished and exploited workers, including tobacco workers, who often receive less than minimum wage. Women work long hours lifting large quantities of tobacco while breathing in thick tobacco dust. They suffer respiratory problems, physical fatigue, and back ailments.

Employers exert a great deal of control deducting wages arbitrarily, often citing loans to employees. Women are disempowered due to (i) poverty, which makes them dependent on their employers for work; (ii) social distance between women and employers from different castes; and (iii) inequality in gender relations. The Self-employed Women's Association (SEWA), now with over 14,500 members, has been striving to improve working conditions for women through unionization.

For example, it has 250 members in Rasnol, the first village it unionized. Rasnol also has a childcare center run by SEWA, a savings group of 50 members, and over

70 women members in an insurance scheme. In Rasnol, as in other areas, SEWA first conducted a socioeconomic survey to give organizers an idea of the problems women faced. Workers' education classes were held to make women aware of their problems and likely solutions. Although SEWA's unionization process is longer than usual, it encourages the workers themselves to take action.

Unionization helps forge a bond among women. Workers' education classes not only made women conscious of their rights, but they also served as a meeting place to exchange views and discuss common problems. In unionized villages, wages rose, the minimum wage is often assured, and issues such as migrant workers and piece-rate overtime-rate wages are dealt with through the proper channels. Beyond trade unionism, SEWA supports childcare cooperatives, savings groups, health workers' cooperatives, insurance schemes, and employment programs.

Nepal is suffering increasing environmental problems. Nepal Forum of Environmental Journalists (NEFEJ) in 1986 was granted NGO status only in 1990, seeks to sensitize journalists, politicians, policy- and decision makers, opinion leaders, and the general public to environmental and development issues in Nepal through the use of radio, television, and print media. NEFEJ to become Nepal's most active NGO. It has succeeded in convincing various political parties that they should include environmental issues in their platforms. In 1992, NEFEJ prepared the Environmental Communication Plan for inclusion in the Government's five-year plan. To help monitor related environment issues, NEFEJ established watch groups for pesticides, wildlife, public health, urban environment, and health.

Civic Environmentalism

Sustainable economic development is one of the most consequential political developments of the past two decades has been the emergence of widespread environmental concern and associated environmental movements throughout the developing world. Framing the core challenge of sustainable development

Civic environmentalism is on the rise throughout the developing world; and the challenge is to adopt innovative policies and institutions that tap this well of social energy and translate public demands and enthusiasm into specific gains in social welfare and environmental quality.

Market power is a subset of people power. In Thailand, environmentalist Buddhist monks have joined forces with other citizens' organizations to prevent deforestation in protected areas. In Tehran, local artists recently organized an exhibition to publicize the plight of children suffering the effects of poor air quality. In Ghana, the League of Environmental Journalists promotes public awareness by working for more effective media coverage of environmental issues. In the Philippines, the nongovernmental Green Forum launched its Earthvote Project in 1991, producing voter guides assessing the environmental credentials of candidates for public office.

In Taiwan, Bolivia, and elsewhere, environmental groups have successfully lobbied for the creation of new environmental laws and regulatory frameworks. Over the past fifteen years in particular, countless numbers of citizens in developing and transition economies have mobilized to protect environmental quality and promote sustainable development strategies.

The advantages of civic involvement in the provision of public services is widely recognized in the literature on NGOs, which emphasizes among other traits their penchant for flexibility and innovation and their role in providing independent assessments and scrutiny of government actions. While the rationale and ethical imperative for citizen involvement in policymaking are widely appreciated, civic engagement takes on added importance in the environmental arena.

At the global level, environmental problems differ in important respects from the concerns that have inspired coordinated international responses in the past. International law typically assumes the form of agreements among governments to change government behaviour, in areas such as nuclear testing and trade policy. In contrast, international environmental treaties are typically agreements among governments to change private behaviour within their borders.

Particularly for global environmental problems driven by diffuse private actions, such as climate change and biodiversity loss, governments cannot effectively undertake such a task without the active engagement of civil society. At the domestic level, the information needed for effective environmental management is often geographically dispersed and location specific. Governments have neither the administrative capacity to collect all of the pertinent information nor the power to prevent public scrutiny of environmental quality.

In contrast to policy decisions regarding exchange rates and military preparedness, for which the relevant information and expertise rest with central planners, citizens can provide high quality information on ecological and human health and on the social behaviours affecting them. This represents a substantial resource for state officials and provides an opportunity and rationale for citizen involvement in environmental management.

In the United States, for example, the Breeding Bird Survey enlists the help of recreational bird watchers in monitoring the status of avifauna by distributing standardized forms for data collection on the distribution and abundance of various species. Civil society organizations also provide valuable information on resource scarcity, particularly when scarcity is not adequately communicated through market prices.

Efforts by university scientists and environmental NGOs to raise public awareness of pending global climate change provides an example of this information mechanism operating on a large scale. The creation and strengthening of national environmental policies and institutions—such as environment ministries, watershed management districts, national parks, and air pollution regulations—typically unfolds over time horizons of a decade or more. Their success depends, in turn, on the long-term presence of policy reformers who can engage in policy experimentation and lesson- learning, take advantage of rare windows of political opportunity, and undertake a cumulative effort - across projects and across administrations - that foreign experts, advocates, and donor organizations cannot.

A civic institutional infrastructure comprised of diverse nongovernmental organizations allows policy reformers to stay involved over the long haul, as these civic organizations provide employment, networking opportunities, and venues for sustained intellectual creation and influence during periods when partisan political shifts prevent their direct participation in government. Public scrutiny of government actions can improve accountability and reduce incentives for corruption.

The wasteful natural resource policies are often the result of efforts by government officials to provide off-budget funding for politically controversial projects. In Indonesia in the 1960s and early 1970s, for example, oil rents normally destined for the national treasury were diverted through the state-run oil company, which President Suharto used to fund initiatives opposed by officials in the planning and finance ministries. Additionally, among economic activities in the developing world, the commercial exploitation of natural resources is particularly susceptible to corruption.

In Bolivia in the mid-1980s, for example, a voracious illegal trade in endangered species was facilitated by customs officials who doctored export permits in the absence of any independent oversight. When citizens have access to independent sources of information, meaningful channels for political participation, and legal protection against retribution, they can comprise a political force for improvements in the performance of public agencies.

The existence of an active nongovernmental sector and associated employment opportunities also make it easier for environmental officials to resign from government service when they deem it expedient, rather than face a choice between unemployment and collusion in misguided policies. This "exit" option gives principled public servants a powerful tool for publicizing mismanagement and for bringing pressure to bear on natural resource agencies loath to undertake needed reforms. The new environmental regulatory agencies that have sprung up across the developing world in recent years are almost invariably less politically powerful than well-established ministries of planning, finance, and agriculture.

Moreover, environmentally-conscious officials working within the traditional ministries must often take a back seat to officials in charge of conventional resource extraction activities. Civic environmental movements lend political support, legitimacy, and expertise to government reformers struggling against conventional bureaucratic approaches and resisting the political demands of those wit h a vested interest in unsustainable resource extraction.

Civic environmentalism frequently catalyses civic engagement on behalf of other social goals. Research on democratic practice shows that meaningful political participation increases the willingness and ability of citizens to organize on issues other than those inspiring their initial efforts. Civic environmentalism has frequently manifested these spillover effects, lending support to indigenous movements in Latin America and democratization in Asia and Eastern Europe. Conversely, social mobilization inspired by other issues (nationalism, racial equality, religion, farmers' rights, women's empowerment) frequently catalyses and shapes the content of civic environmentalism.

The advantage of the potential synergies between civil society activity and good governance, we must move beyond analytic frameworks that portray state-society relations as a zero-sum contest. It has become fashionable in many circles to interpret the recent strengthening of the private and non-profit sectors as tantamount to a decline in the power, relevance, and responsibilities of government. But this interpretation of the relation between state power and societal strength rests on both conceptual and empirical flaws.

Conceptually, the problem arises in the failure to distinguish between two types of power: monopoly control and the capacity for action. Viewed through the former lens, the empowerment of civil society represents a loss in government power, by definition.

Trends such as the decline of state socialism and the expanding scope and activities of nongovernmental organizations do indeed represent a weakening of government monopoly power in cases where state agencies dominated sectors of economic activity and service provision. But if we recast the concept of state power as the capacity for governance, and think in terms of absolute rather than relative resources, a different picture emerges.

When expanding markets provide a larger tax base, or partnering with NGOs increases the level and efficiency of government service provision, in absolute terms the capacity of the state is enhanced. It is well established in the literature on democracy that an engaged civil society can strengthen the ability of government to provide public services, including in the environmental realm. When a business divests itself of operations that detract from its core strengths, or cuts its workforce to balance the books, it is typically judged by industry analysts as having become stronger.

In contrast, when a government divests itself of unprofitable industries that it is ill-equipped to manage, or cuts back bloated bureaucracies, commentators frequently proclaim that state strength has diminished.

Again this perspective conflates the concept of state power with that of monopoly power. Even if state power is construed narrowly in terms of a government's ability to acquire resources, monopoly control is only one way to achieve this goal and often an ineffective one at that.

All forms of cooperation are in a very real sense arrangements in which actors sacrifice a measure of control in exchange for greater benefits than they might accrue acting alone. Just as a firm entering into a business partnership relinquishes some control in exchange for greater long-term profits, when a national park agency works in partnership with NGOs it can create a larger and more ambitious protected areas network than it could by relying on agency resources alone.

Government authorities can then focus on their core competencies in areas such as law enforcement and land use planning. A decline in monopoly power is often a prerequisite for an increase in the capacity to govern. Empirically, the historical and cross-national evidence show that the rise of nonprofit organizations is not generally associated with reductions in state spending.

There are numerous instances in which state authorities and civil society have been engaged in a mutually hostile relationship whereby a gain for one is perceived by both sides as a loss for the other. This is why civic environmentalism represents an ideological and administrative challenge to that relationship, requiring new institutional approaches that can overcome a legacy of mistrust and take advantage of the potential for joint gains.

In India and Indonesia, for example, reforms in government forestry agencies are designed to improve community relations and reverse a pattern of conflict, inherited from the colonial period, in which forest police forcefully excluded communities from traditional resource extraction activities. In many parts of Latin America, indigenous groups whose demands have long been ignored by governments are for the first time gaining land rights and management authority in protected areas to promote the twin goals of self-determination and environmental stewardship. These experiences are part of a larger trend, as the zero-sum paradigm has increasingly been called into question by social movement participants in developing countries.

In nations that have recently undergone a process of democratization, activists who built their skills and credentials as opponents of authoritarian regimes are assuming new roles and seeking new forms of collaboration with government authorities. Throughout Latin America, for example, nongovernmental organizations are working with municipal authorities to improve the delivery of services to local communities.

Civic Environmentalism in Developing Countries

Environmental management are now a regular feature of political discourse throughout the developing world. Over the past fifteen years in particular there has been a surge of public interest and action in this area. Today the level of citizen participation in environmental groups in the developing world compares.

The rise of civic environmentalism has taken place in a broader context of increasing civic organization throughout the developing world. Non-environmental organizations include professional, religious, artistic, charity, and sports associations, as well as political parties. In addition to the direct causal connection between environmentalism and other forms of civic engagement, the across-the-board increase in social organization is indicative of systemic trends simultaneously affecting multiple facets of society.

In many countries significant environmental movements first arose in the 1980s in the midst of broader social movements for democratization. The causal relation between democratization and environmentalism has never followed a simple formula, but has unfolded in diverse and historically contingent ways in different countries.

In South Korea, social movements for democratization and environmental protection joined forces in opposition to authoritarian rule in the 1980s, and many leaders of the democracy movement later assumed leadership positions in the country's environmental organizations following the consolidation of a democratic regime in 1992. A similar pattern unfolded in Taiwan, where the environmental and pro-democracy movements were the two strongest social mobilizations opposing the KMT's monopoly on political power.

In Brazil, disparate environmental organizations that had maintained a low profile during military rule were united and animated when they helped draft the environmental chapter of the new national constitution in 1985-1988. The resulting alliances formed the basis for coordinated action among local and regional groups in subsequent years. In the Soviet Union, civic environmental organizations flourished in the early years of the USSR, were crushed under Stalin, resurfaced in limited form during the political liberalization of the

1950s, and exploded on the scene as a central component of mass movements for democratization in the late 1980s.

The rise of civic environmentalism in the developing world and Central and Eastern Europe has also been fostered by international trends, including an increase in cross-border organizing by environmental groups. In the United States, the Reagan administration's efforts to roll back environmental regulations in the 1980s sparked a public reaction that contributed to tremendous growth in the size and resources of U. S. environmental groups.

Concurrently, growing concern over global-scale problems such as climate change, ozone depletion and tropical deforestation motivated these groups and their Western European counterparts to significantly expand their international programs, and they soon committed millions of dollars to support environment efforts in developing countries. Combined with a growing interest in environmental issues on the part of official development agencies and the increasing tendency to funnel ODA through nongovernmental organizations, the 1980s represented a watershed with respect to the resources available for civic environmental organizing in developing countries.

The cumulative impact of these international resources is substantial: In Costa Rica and Bolivia, most of the funds for the major environmental organizations come from abroad. In India, roughly 90 percent of all funds for nongovernmental organizations is of foreign origin. According to the 1998 Global Environmental Organizations Survey, which polled 248 environmental groups in 59 countries, 45 percent of all groups report that they have contact with international NGOs "very often".

Some study suggests that the political challenge of sustainable development encompasses more than the widely-recognized imperative of cross-sectoral coordination among government agencies. Mobilizing the energies of civil society for sustainability means incorporating these concerns into the mandates of traditional civic institutions such as religious and cultural organizations, day care centers, health service providers, private schools, and recreation clubs. In practice, cross-sectoral connections within civil society have played an important political role in countries experiencing large-scale social mobilization for environment causes.

The involvement of diverse civic organizations in environmental movements enhances the credibility and visibility of environmental concerns, broadens the base of political support, and helps to ensure that environmental ideas are interpreted in a manner that reflects the unique history, culture, and social and physical conditions of a particular country, rather than resembling a crude copy of foreign ideas and movements. This cross-sectoral dynamic can be seen in Bolivia, where the women's rights organization Casa de la Mujer runs two-week training sessions for migrant workers that combine education on health, sexuality, legal rights, and environmental management.

In the Bolivian Chaco region, environmental and indigenous groups have joined forces to create innovative park management schemes granting land rights and administrative authority to the Izoceño-Guaraní. In other instances, potential opportunities for cross-sectoral alliances have been lost; Bolivian environmentalists have yet to take up the occupational health hazards of mining activity as an environmental cause, which could forge important links with powerful labor unions. While there is significant social concern for the environment in many developing countries, this concern exists alongside high levels of air pollution, inadequate access to drinking water, widespread pesticide poisonings, and mismanagement of natural resources.

As a rule, environmental outcomes in developing countries fall far short of expressed public preferences. Given the importance of both government policy and social behaviour in shaping these outcomes, the next section considers a suite of measures that can be adopted to improve the effectiveness of environmental management through innovate state-society collaborations.

Civil society can exercise a powerful influence on the ability of government to provide public goods, the development, autonomy, and impact of civil society organizations is strongly shaped by government policies. The Global Environmental Organizations Survey reveals that environmental groups devote a considerable portion of their energy to interactions with government authorities.

Civic environmentalism is most able to flourish where democracy is strongest. The generic challenges of sustained collective action are compounded when citizens fear for their safety or operate in a political environment where autonomous civic organization and the expression of dissenting views are considered a threat by state authorities. The protection of civil liberties extends beyond protection from abuses of power by government. Environmental problems are often the result of rent-seeking and externality-producing behaviour by private interests who stand to lose large sums if environmental advocates have their way, and there is a growing list of environmentalists in developing countries who have been harassed, imprisoned or killed at the behest of affected interests.

Civic environmentalism requires that citizens enjoy active protection by the state against intimidation or retribution by other social actors. Judges and law enforcement authorities must also receive training to familiarize them with the content of legislation for environmental protection and citizen rights, so that they can exercise their responsibilities in accordance with the law. Autonomous media organizations play an important role in promoting civic engagement generally and civic environmentalism in particular. A free press can provide independent sources of information and assessment regarding trends in environmental quality and on the role of government and private actors in shaping these trends.

Media organizations often help to foster the growth of environmental movements by publicizing the actions and accomplishments of isolated citizens' groups. Print media also provide an essential outlet for public intellectuals engaging in the crucial task of shaping environmentalism into a philosophy that resonates with national needs and aspirations. The close relation between civil liberties and civic environmentalism can be seen in the growing role of indigenous organizations in environmental management.

To be effective land managers, indigenous groups require rights to land and resources, recognition as legitimate political actors in their countries, and equal protection under the law. The expanding involvement of indigenous groups in national park management has proceeded in lock-step with advances in their civil liberties and their growing prominence in national politics. Central government authorities can also take a proactive role in protecting political liberties by requiring local governments to create mechanisms for citizen participation as a prerequisite for administrative decentralization and revenue sharing.

In the Philippines, passage of the Local Government Code in 1991 has created new opportunities for participation by civil society on local development councils and resource management councils. In contrast, recent decentralization efforts in Lebanon and in the West African Sahel have often been implemented without institutional mechanisms to ensure broad participation. Such initiatives run the risk of exacerbating local disparities in resource access, which are often maintained by entrenched local interests operating through institutions inherited from the colonial period.

A comprehensive empirical analysis of political participation in urban settings in the United States revealed that local democratic participation is strongest in cities where the federal government mandated the creation of participatory institutions as a prerequisite for federal block grants. Governments can also enact legislation affording special protection to "whistle blowers"—individuals in government agencies or businesses who speak out against corrupt, illegal, or otherwise harmful practices by their employers. These individuals are privy to information that is difficult for members of the media or government watchdog agencies to acquire, and can play a valuable role in preventing and reporting environmental crimes.

Whistleblower legislation is a prime example of the need to create an institutional environment that mobilizes the energies of citizens for the sake of environmental management. For an employee of a mining or timber company to report an environmental crime, this person must not only possess a personal concern for the environment but must also have the benefit of legal institutions that allow the employee to translate that concern into environmental improvements. Civic environmentalism can also be promoted through electoral reforms that establish regular opportunities for voter referenda on specific policy issues, as is common practice in Switzerland.

Many nations invite nongovernmental representatives to serve as advisors and even official delegates at international environmental negotiations. This is a particularly useful strategy for small countries that lack the capacity to field the large numbers of government delegates needed to effectively engage in complex negotiations. Conversely, government agencies can provide incentives to encourage civil servants to spend time working in nongovernmental organizations. A variety of additional measures can be adopted to encourage state-society collaborations.

The Philippine Department of Agriculture created an NGO Outreach Desk that has strengthened relations with the nonprofit sector and has facilitated collaboration on projects ranging from sustainable inland fisheries to the application of sloping agriculture technologies for the prevention of soil erosion. In Bolivia, national park authorities contract out park management to a variety of nonprofit and indigenous peoples' organizations.

In 1999 Brazil's government to contract NGOs to carry out functions such as park management, educational campaigns, and research for government agencies. Governments can also provide training and employment opportunities for "barefoot environmental managers." In Costa Rica, for example, "parataxonomists" from rural communities receive basic training in the collection and preparation of specimens for taxonomic identification by scientists at the National Biodiversity Institute (INBio).

In South Africa, the government has launched a highly successful Working for Water Programme, which employs thousands of disadvantaged citizens in a program to control invasive plant species that threaten the country's scarce water supplies. Government agencies frequently have the ability to amplify the impact of civil society innovations in areas like agroecology by incorporating these into standard agency practices, publicizing innovations to other user groups, or implementing them on a larger scale. While the small scale and flexibility of most NGOs is conducive to creative problem-solving, some types of innovation require large-scale research activities.

Agricultural centers of excellence under the CGIAR system, for example, excel by bringing together large numbers of prominent scientists to create a critical mass conducive to major research breakthroughs. In these circumstances, just as governments are well-suited to scaling up local innovations, NGOs can help to "scale down" technological innovations developed with broad national or regional conditions in mind, tailoring them to the specific needs of local user groups.

Although environmental problems are often described in catastrophic terms - and occasionally conform to such descriptions, as when droughts, floods, or famines occur - many environmental problems are difficult for ordinary citizens to detect. This is the case, for example, with DDT in breast milk, odorless toxic emissions from factories, or the quiet decline of a rare plant species in a remote location. One of the most powerful tools that governments can use to catalyze civic environmentalism is to publicize information on environmental quality in venues readily accessible by the public and in formats that can be understood by non-experts. When government agencies publicize air quality data on a routine basis, as is done in Taiwan, they help to create a social constituency in support of government regulatory efforts.

In 1995 environmental regulators in Indonesia launched the widely-acclaimed Program for Pollution Control, Evaluation, and Rating (PROPER), which produces color-coded ratings to signify the extent of private firms' compliance with water pollution laws. The rankings are publicized using methods designed to maximize the reputational pressure brought to bear on polluting industries. Combining public and market pressures, the program has resulted in improved compliance and is considered a significant success. PROPER is another example of how innovative institutional arrangements can help translate a public's general unease regarding pollution into specific demands that can serve as the basis for improvements in environmental quality.

Environmental impact assessments have the potential to inspire significant levels of citizen participation. Alternatively, they may amount to little more than planning documents for small groups of specialists. Government agencies should adopt policies to promote the former approach, creating a meaningful role for groups potentially affected by proposed projects.

Planning authorities should involve citizen stakeholders early in the process to ensure that they are not merely commenting on the report's conclusions, but providing input into the types of questions to be addressed.

The initial impact assessment should be widely publicized and followed by an adequate period for public review and comment. Public comments can then be published in the revised report together with replies from the responsible government agencies. Information provision is most powerful when combined with measures to ensure that citizens can act on that information. When citizens are granted legal standing to help enforce environmental laws, the information in environmental impact assessments can serve as a basis for grievances brought before the courts. Governments can also make high quality environmental information available to the public on a proactive basis, rather than solely in response to proposed projects or polluting activities. One approach would be to establish national environmental depository libraries.

Environmental depositories could serve as national clearinghouses for the wealth of environmental information routinely sent to regulatory authorities by foreign and international environmental agencies. Authoritative assessments of environmental issues too often end up in the private collections of government officials or are dispersed among reading rooms of NGOs and issue-specific scientific institutes. Situated in existing, well-known libraries, environmental depositories could hold publications from energy, agriculture, and transportation ministries, student theses, U. N. reports, data from nongovernmental organizations, and national and foreign environmental impact assessments. The latter contain a wealth of information regarding specific industries and describe impacts and mitigation measures (including best practices) commonly associated with particular types of developments.

Important decisions must be made about the types of publications to place in environmental depositories; the goal should not be to replicate existing collections, but to provide concerned citizens with convenient one-stop access to environmental information. Finally, "freedom of information" legislation requires government agencies to provide information requested by citizens within a specified period of time. These must be implemented in such a way as to maximize the quantity and quality of information made available to the public while minimizing the associated administrative burden.

Public access to information is guaranteed in the South African constitution and is the focus of legislation in a growing number of developing and industrialized nations. More recently public access to information has become a focus of global advocacy campaigns and has been encoded in international law in the Aarhus Convention.

Governments can provide reductions in personal and corporate taxes for those who contribute money, land, or other resources to nonprofit organizations. These incentives have played an important role in the development of the nonprofit sector in many countries. One recent example is the striking growth of private land conservation organizations, or "land trusts," in the United States.

Land trusts typically operate at the community level and are therefore able to craft arrangements in a cooperative manner with landowners who might be wary of government restrictions on private property. The nonprofit organizations brokering these deals can act quickly when land becomes available for sale, enabling them to take advantage of market opportunities that would be lost if one had to wait for a government land conservation agency to clear the necessary bureaucratic hurdles.

Land trusts often subsequently sell the title to the appropriate government agency once the agency has obtained the necessary funds and administrative authorization. This allows the agency to specialize in its comparative strengths in revenue generation and land management, while land trusts provide flexibility and the local knowledge and credibility needed to proactively identify opportunities and broker deals.

Tax incentives of this nature may be difficult to use in countries that lack a reliable revenue collection system or rely primarily on sales tax. As an alternative, local governments can solicit voter consent to create special tax assessment districts in which a small percentage of sales or property tax is devoted to the purchase of species habitat, recreation areas, watersheds, or farmlands in rapidly urbanizing regions whose citizens wish to preserve their agricultural heritage. Governments must also set standards that specify qualifications for nonprofit status.

Absent such standards, organizations lacking a public service mandate may portray themselves as nonprofits with the aim of generating revenues or reducing tax burdens. It then becomes more difficult for legitimate organizations to attract scarce resources. In a similar vein, government agencies can either create or grant official recognition to labelling and certification programs for environment- friendly products and services, such as organic agriculture, ecotourism, and sustainably- harvested timber.

In the absence of certification standards, conventional businesses may attempt to appropriate environmental symbols and rhetoric without changing practices, which removes the incentive for innovative firms to invest the extra time, energy, and resources needed to develop environmentally superior approaches. Governments must also devise regulatory frameworks to guide the practices of nonprofit organizations.

Although civil society organizations are frequently a cause and consequence of large-scale democratization, they are themselves only rarely democratically accountable beyond their client base, membership, or financial supporters. A measure of accountability is necessary when they receive public resources for public purposes. A balance must obviously be struck between the need for NGO accountability and freedom from undue government interference.

Environmental funds serve as coordinating mechanisms for attracting and managing foreign environmental aid and offer a valuable service to civic organizations by reducing the administrative burden associated with fund-raising. One of the great attractions of environmentalism as a political cause is the opportunity that it provides for hands-on involvement at the local level.

By increasing opportunities for meaningful participation, government authorities can help spawn volunteerism and autonomous coordinated action by concerned communities. In many countries national park authorities invite youth organizations to participate in trail-building activities.

Communities can also be encouraged to participate in adopt-a-creek programs, taking responsibility for monitoring the condition of local watersheds. In 2000, the annual International Coastal Cleanup inspired volunteer efforts in 73 countries; in the Philippines, organizers reported that over half a million volunteers collected 6.3 million pounds of garbage.

In many countries, regulatory agencies and citizens' groups have created environmental complaint hotlines to encourage citizens to report environmental crimes. In Costa Rica, the Ministry of Justice provides badges to volunteers from the nongovernmental group APREFLOFAS, which organizes citizen patrols to report illegal extraction activities in national parks. Governments can adopt additional measures to solicit input from citizens when designing and implementing environmental policies.

Many countries create public commissions and discussion forums prior to major international environmental negotiations, as well as permanent local management committees for communities living in or near protected areas. It is essential that local environmental management committees are granted real decision- making or budgetary authority, or otherwise provide significant benefits to group members to justify the cost of participation.

During the year 1970s, environmental advocates in many developing countries were derided as partisans of a fringe concern bearing little relation to mainstream development needs. The mainstreaming of environmental concerns has resulted in part from efforts by political leaders in the South to embrace and make public pronouncements on behalf of sustainable development.

This legitimizing function of government remains an important means by which state authorities can foster and support civic environmentalism at home.

Governments can also complement the efforts of NGOs and news media to raise awareness among the general public, by incorporating environmental themes into national education curricula and standardized exams that influence the teaching practices of state and private institutions. Other environmental education activities include nature interpretation services within national parks, which help the general public to understand and value their nation's biological heritage. Governments can also highlight outstanding instances of civic environmentalism by providing public recognition for citizen accomplishments in environmental research, education, advocacy, and management.

Political participation demonstrates that recruitment to political causes occurs through social networks in which members tend to recruit people who share their racial, gender, and class characteristics. Governments can help to both broaden civic environmentalism beyond the community of environmental specialists and deepen it to include not only society's most active and politically engaged civic leaders, but the general citizenry as well. As noted earlier, one of the core challenges of civic environmentalism is to establish cross-sectoral linkages among diverse nongovernmental organizations promoting causes such as community health care, rural development, recreation, religion, and the arts.

National governments are well-suited to help with this process because they maintain relations with a much wider variety of interests and constituencies than do leaders of environmental organizations. Ministries of sports and youth can encourage sports clubs and hunting and fishing organizations to become involved in environmental activities. Health ministries can arrange for hospitals and private clinics to collect environmental health statistics on acute pesticide poisonings and respiratory diseases exacerbated by poor air quality.

Ministries of art and culture can sponsor environmental art competitions, exhibitions, and public works to enlist a nation's most respected cultural figures in communicating environmental themes. In the Seychelles, the Ministries of Education and Tourism co-sponsor an annual festival on underwater photography that engages students and teachers in activities designed to raise awareness about coastal and marine issues. When contracting with environmental nonprofits for the provision of services such as agricultural extension, pollution monitoring, or park management, governments can stipulate that services be provided in such a way as to include groups or communities that environmental NGOs might otherwise miss.

In the course of environmental impact assessments, government agencies can take steps to ensure the participation of all relevant actors rather than hold closed consultations with leaders of prominent environmental organizations. Beyond questions of breadth, governments can make efforts to deepen the engagement of civil society, involving those who lack either the resources or inclination to assume a leadership role.

As a rule, people with higher levels of income and education participate more extensively in social organizations and political life than do those with fewer resources. To the extent that governments can find ways to lower the barriers to participation for poor and marginalized people—through schools activities, for example—civic environmentalism will have assumed a deeper shade of green.

Sustainable Livelihood

A sustainable livelihoods approach suggests that helps to bring together different perspectives on poverty and integrate the contributions to eliminating that poverty different skills and sectors can make, in for instance designing projects and programmes, sector analysis and monitoring; makes explicit the choices and possible trade-offs in planning and executing different development activities; helps to identify the underlying contraints to improved livelihoods and the means of overcoming these; helps to link improved micro-level understanding of poverty into policy and institutional change processes.

Poverty has most commonly been assessed against income or consumption criteria. Dissatisfaction with the income/consumption model gave rise to basic needs perspectives which go far beyond income, and include the need for basic health and education, clean water and other services which are required to prevent people from falling into poverty. More recently, poverty has been defined in terms of the absence of basic capabilities to meet these physical needs, but also to achieve goals of participating in the life of the community and influencing decision-taking. A sustainable livelihoods (SL) approach draws on this improved understanding of poverty, but also on other streams of analysis, relating for instance to households, gender, governance and farming systems, bringing together relevant concepts to allow poverty to be understood more holistically.

The SL approach supports poverty eradication by making enhancement of poor people's livelihoods a central goal of development efforts.

A livelihood is sustainable when it can cope with and recover from stresses and shocks, and maintain or enhance its capabilities and assets both now and in the future, while not undermining the natural resource base'. The assumption is that people pursue a range of livelihood outcomes (health, income, reduced vulnerability, etc.) by drawing on a range of assets to pursue a variety of activities. The activities they adopt and the way they reinvest in asset-building are driven in part by their own preferences and priorities.

However, they are also influenced by the types of vulnerability, including shocks (such as drought), overall trends (in, for instance, resource stocks) and seasonal variations. Options are also determined by the structures (such as the roles of government or of the private sector) and processes (such as institutional, policy and cultural factors) which people face. In aggregate, there conditions determine their access to assets and livelihood opportunities, and the way in which these can be converted into outcomes.

Thus, the poor may draw on social capital such as family or neighbourhood security mechanisms at times when financial capital is in short supply. The people may, for instance, depend on their own farming, on selling their labour locally, or on migration, all within the same year. Outcomes will not be simply monetary, nor even tangible in all cases. They may include, for instance, a sense of being empowered to make wider, or clearer, choices.

For instance, a household's long-term strategy may be to reduce its vulnerability to drought, and reducing rainwater run-off from farmland may be one set of activities within this strategy. However, some of the necessary labour may be diverted in response to new migration opportunities, or some of the capital needed may be diverted to respond to a medical crisis. In practice, an SL approach is therefore essentially concerned with the dynamic and, at times, iterative nature of livelihood strategies.

A livelihoods approach puts people at the centre of development. This is equally important at macro levels (e.g. in relation to economic reform) as it is at the micro or community level. This means that practical applications of SL concepts:

— start with an analysis of people's livelihoods and how these have been changing over time;

— fully involve people and support them in achieving their own livelihood goals;

— focus on the impact of different policy and institutional arrangements on people's livelihoods; and,

— seek to influence these arrangements so they promote the agenda of the poor.

SL concepts allow the identification of livelihood-related opportunities and constraints regardless of where these occur:

— it is non-sectoral and applicable across social groups;

— it recognises multiple influences on people, and seeks to understand the relationships between these influences;

— it recognises multiple actors (from the private sector to national ministries, from community-based organisations to newly emerging decentralised government bodies);

— it acknowledges the multiple livelihood strategies that people adopt to secure their livelihoods;

— it seeks to achieve multiple livelihood outcomes, to be determined and negotiated by people themselves.

In 1960s and 1970s, the ideas of Integrated Rural Development Programmes , where external interventions can best be made are formed holistically. But by contrast, the implementation of these need not be multi-sectoral: SL approaches recognise that support can best be initiated in response to particular opportunities or needs, even within sub-sectors or within small areas, and then gradually expanded.

However, many factors that affect livelihoods have distinctly macro characteristics. For instance, natural capital may be threatened by flash-flooding, which will be influenced by the design and implementation of policies to prevent deforestation further upstream; access to financial capital will be influenced by policies towards credit and the rural banking sector more generally; vulnerability may increase or decrease depending on, for instance, on how well emergency feeding and employment schemes are designed and implemented.

The Africa Wildlife Foundation assessed the economic and livelihood impacts of wildlife enterprises in Kenya. This moved beyond financial analysis, and sought to understand long-term developmental impact Programme review. A review of community-based natural resource management (CBNRM) in Namibia sought to synthesise information on livelihood strategies, how they affect people's participation in CBNRM and how CBNRM can best 'fit' with livelihoods.

A wider view is that sustainability is achieved when overall stocks of physical capital (in whatever combination) are maintained. This can allow, for instance, for losses in biodiversity as modern crop varieties are introduced. Sustainability relates not just to natural resources, nor to stocks of physical capital.

For specific activities to be maintained, the institutions underpinning them—whether traditional, governmental or commercial - need to be sustained. This would imply, for instance, that state subsidies or any other intervention in response to market failure should be sustainable. But even this does not fully capture 'sustainability' for, as we have argued, portfolios of activities shift in response to people's capacity to generate new activities in response to needs and opportunities.

As people develop their capacity to switch among activities, they may, as argued above, substitute one type of capital for another. The feasibility or acceptability of interchanging types of capital will depend on the context in which people live. They may also face wider tensions, including those between:

— maximising short-term income or guarding against vulnerability, and responding to concerns about longer-term environmental sustainability; and between

— the achievement of individual, household or community livelihood objectives and the desirability of not compromising livelihood opportunities open to others.

The livelihoods were less dependent on natural resources (NR) than expected, partly because the poor had such limited access to these resources. An SL approach helped to identify how to empower the poor to break existing cycles of impoverishment, by strengthening their access to capital assets, and by working at policy and institutional levels to strengthen the capacity of the public sector, local government and NGOs jointly to design and implement poverty-focused programmes.

A livelihood analysis of butterfly farming in Kenya An internal review of a Kenyan butterfly-farming project in 1997 asked farmers how much they earned from butterflies and from agriculture. The analysis of livelihood impacts in 1998 was much more finely textured, identifying improvements in income and support to household strategies of diversification, but less improvement in security because earnings were unpredictable and the activity is risky. The high level of risk explains low participation among poorer farmers. Project impacts were assessed in relation to all elements of the livelihood framework. This highlighted positive and negative affects on assets, and two other advantages: the low trade-off with other activities, and increased access of butterfly farmers to external institutions, such as the Forest Department.

In Namibia, in the early 1990s a Community-Based Natural Resource Management Programme (CBNRM) focused mainly on institution-building and the legal rights required for cash incomes to be generated. By 1997, field staff had gained insights into livelihood issues affecting, and affected by CBNRM, while agricultural staff had accumulated complementary information on agricultural strategies, through Farming Systems Research and surveys.

For poorer households, access to small amounts of income, elephant damage to crops, and continued access to wildlife/tourism areas for harvesting plant resources were the critical issues. A large minority faced a considerable seasonal cash shortage, so even small amounts of CBNRM income could help meet food needs or school bills if timed correctly. The analysis highlighted that minimising costs to livelihoods was as important as maximising benefits.

In Namibia, the Wildlife Integration for Livelihood Diversification (WILD) project used a wide range of participatory techniques to assess how different project options affected, or were affected by people's livelihood choices.

Learning processes are important to promote on at least two levels: first, the teams using the approach need an initial familiarisation and learning period, where necessary devising simplified versions of the framework for local presentation.

In Orissa, concerns to identify the livelihood needs of the most vulnerable, initiate social organisation and capacity building, and facilitate negotiation of their resource rights led to the establishment of Livelihood Support Teams, comprising government and NGO staff, and individuals from Orissa and beyond.

Poverty elimination is not only a matter of supporting local-level initiatives in response to needs and opportunities which the poor perceive. Many facets of their situation which they may perceive as 'fixed' (such as, for instance, weak provision of health and education services) may, in fact, be amenable to change.

2

Environment and Development

Institutions for sustainable development are to protect people and a broad portfolio of assets, they must respond to and shape the major changes that will unfold over the next 50 years: urbanization, technological innovation, economic growth, shifting social values, changing scarcities for environmental and natural assets, and stronger linkages among nations.

Institutions thus must be stable, but they also must be capable of changing and adapting, and new institutions must emerge. Institutions are the rules and organizations, including informal norms, that co-ordinate human behaviour. They are essential for sustainable and equitable development. When they function well, they enable people to work with each other to plan a future for themselves, their families, and their larger communities. But when they are weak or unjust, the result is mistrust and uncertainty.

Sometimes social and economic development offers opportunities for change. Structural changes-urbanization, the demographic transition, the redistribution of wealth-unleash dynamic forces and opportunities for institutional change. Initiatives to channel information can also serve as catalysts for change. Information can empower people by giving them more voice in public services and allowing greater transparency and accountability in the activities of governments and firms. Institutions are the rules, organizations, and social norms that facilitate coordination of human action.

The distinction between social capital and institutions can sometimes be blurred, and there are strong influences between the various social assets. For example, the exchange of goods and services may be based on personal networks and other forms of social capital in the village, but on formal institutions in the city. Similarly, general trustworthiness in a society can be strong either because of strong personal networks-or because of good laws and judicial systems that are generally accepted.

In fact, as societies become more complex, trust in individuals (based on knowledge of character and frequency of interpersonal contacts) is supplemented by trust in institutions when dealing with strangers. Two very important dimensions of coordination are others and future.

Markets are institutions with coordinating functions. A market co-ordinating the transactions of individuals and firms enables them to serve others and invest for the future (as when a baker builds an oven in response to greater demand). But markets need the support of other institutions to ensure confidence, control, and the right incentives.

Economic agents face transaction costs, and institutions can coordinate to lower those costs. Traders may want to cheat each other-or to renege on their promises. But social capital and modern institutions-such as the rule of law and the enforcement of contracts-reduce this probability, facilitating mutually beneficial transactions. Particular challenges for these other institutions are to commit to protect and nurture people and assets-and to serve dispersed interests.

Firms, government, and civil society are positioned to act and to influence the actions of others, playing complementary roles in coordination. Market participation enables people to specialize and work together, and to apply their skills and resources in the best way possible, as the price mechanism provides information to coordinate. But the institutional framework must ensure that markets function and deliver their expected benefits.

Where formal institutions are weak, activity will locate in the informal sector-in many developing countries the share of workers in the informal sector is more than 60 percent. This means that a substantial proportion of economic activity is deprived of potential productivity-enhancing support, and may also escape guidance from institutions that manage environmental and social aspects. But market power can also give large domestic firms or multinational corporations potential coercive powers akin to the government's in terms of serving narrow rather then broad interests.

Thus, a major challenge for government and formal institutions is to be more welcoming and supportive of private actors with appropriate safeguards to ensure the public interest is not compromised. In many areas, government plays a central role in organizing dispersed interests: meeting national goals and balancing competing interests. Unlike social norms and values, government operates a rulemaking process by which rules can be changed more quickly, with vision and design, and still be forceful.

The private sector will be less willing to invest and do business if instability and risks of expropriatory consequence have not been curtailed. Unless institutions succeed in separating the powers of government and providing meaningful checks and balances, communities and the private sector will be less forward looking, and environmental and natural assets will be hurt through inappropriate investment and conservation. Democratic forms of decisionmaking, despite their limitations, are associated with processes that exhibit desirable institutional features; they pick up signals, balance interests, and execute chosen actions.

Empirical evidence strongly supports Amartya Sen's finding that democracy-helped by free speech-plays a key role in eliminating famine and eliciting effective disaster relief. Theoretical and empirical evidence also suggests that environmental commitment and related outcomes are positively correlated with democratic practices, though some countries have also done well environmentally using other channels. There has been a sharp increase in the active membership of civil society organizations in the past two decades.

During this period civil society organizations have become more capable, sometimes acting independently, sometimes influencing the activities of government and the private sector. Civil society's role has been most notable in mobilizing support for specific issues, supplying information, and providing third-party verification. Civil society organizations often take initiative and provide voice for unheard interests-building the trust, legitimacy, and knowledge needed.

In West Bengal, India, the Ramakrishna Mission works with youth clubs in 1,500 villages, undertaking needs assessments and helping to identify external support for social welfare and development projects. In Thailand the Appropriate Technology Association has facilitated collaborations among farmers, university scientists, and government authorities to reduce pesticide poisoning through the use of biological controls. Civil society and the media also play an important monitoring role-for example, improving accountability and reducing incentives for corruption in government.

The commercial exploitation of natural resources is particularly susceptible to corruption; for example, in Bolivia, a voracious illegal trade in endangered species in the mid-1980s was facilitated by customs officials who, in the absence of independent oversight, doctored export permits. An active civil society makes it easier for environmental officials and others to publicize mismanagement; for example, in Cameroon civil society plays a role in monitoring a promising new logging concession scheme. When citizens have access to independent sources of information, meaningful channels for political participation, and legal protection against retribution, they can become a strong political force for improved performance in public agencies.

Civil society organizations are particularly important for promoting environmentally sound development strategies. Environmental management often is about "downstream" spillover effects, and NGOs can often provide knowledge and links between geographically dispersed locations. The surge in civic environmental activity has often been part of a broad upward trend in civil society activity generally. Opinion surveys consistently reveal levels of public environmental concern in developing countries equal to that in industrial countries.

Institutions that restrain the taking of assets-through norms or under threats of punishment-are essential for assets to thrive, whether the assets are human-made or natural, whether they are transacted in markets or not. For example, restraint is necessary for the forest and the aquifer to yield sustained benefits when population density, changes in technology or preferences, and other developments increase the demands on their use.

Institutions ensuring such restraint allow assets to thrive; however, because they yield dispersed benefits, such institutions do not emerge easily. Protective institutions are those that define and support control rights in terms of access to and use of assets central to human well-being; for example, who can graze their cows where, who controls a factory, who takes home eventual profits-or who is allowed to discharge emissions-when and where, and at what price. A special subset of protective institutions are private property rights. They entail well-delimited (and not too limited) rights of use and decisionmaking for an owner, typically including rights to sell or lease an asset.

These institutions include a commitment from society (extended family, neighbours, villagers, or governments) to help protect these rights. In a modern state, this commitment requires an active obligation of enforcement from the government (police or judges) and the assurance that the government itself respects those rights. For this reason, the security of property rights is closely associated with the rule of law-so that people can make assumptions about what will be respected as theirs. But protective institutions also arise to manage assets that are not amenable to private ownership.

A pollution control agency defines and protects control rights to the air. A central bank protects the integrity of a currency and financial system. And communities can solve coordination problems and allow assets to thrive, as the literature on common property resource management and social capital demonstrates (the important interplay between community cooperation and modern society's formal institutions can be beneficial or detrimental).

In the broader institutional environment, firms, government, and civil society organizations together give force to rules and norms. For instance, civil society can help keep the judiciary independent. Indeed, rules and norms are typically backed by sanctions, and many are effective only when agencies and organizations back them up. Some assets protect other assets-and some are more vulnerable to predation than others.

Land is protected by formal institutions (property rights, courts, and titling agencies) in a modern urban setting, but it is relatively less protected under periodic, nomadic grazing. Fisheries turn from needing no protection when stocks are abundant-to needing costly protective institutions as fishing technology and demand develop. Savings under mattresses need some protection, but those in banks need good vaults and such credible institutions as independent regulatory agencies and central banks. Higher national income can contribute to better institutional quality (for example, through more expenditures on courts). But more importantly, better institutional quality can also contribute to higher national income, as when good institutions facilitate investments or curb overfishing.

A large body of theoretical and empirical studies concludes that there are strong causal effects from good institutions, measured by such variables as rule of law to higher incomes per capita. According to one study, better voice and accountability would raise national income per capita by a factor of 2.5. Another study suggests that per capita incomes would grow at least 2 percent per year in all countries if they would only protect property rights (public and private) and pursue more competitive market policies. So when institutions allow assets to thrive, an economy can flourish.

Valuable assets cannot thrive, and can be in wasteful decline, if there is no social commitment to developing protective institutions. For agricultural land, two facts facilitate the emergence of protective institutions, as they may have come about historically. First, when land becomes scarce, competing users cannot avoid meeting face to face, and can choose to fight or negotiate. Second, when neighbours are settled, they can help each other defend their land in a reciprocal fashion. Therefore, even without a state or feudal lord, transgressors will face some deterrence.

Fisheries account for 19 percent of total human consumption of animal protein, so good institutions for fisheries are important. A typical story of fisheries development starts with the discovery of new fish stock. As fishermen make good money, more capacity rushes in. Though scarcity and wasteful competition ensue, fishermen do not coordinate management. Over time, the catch falls, first per hour of effort, then in total, with individual fishermen barely breaking even. This wasteful race to exploit the resource wastes labour and capital, since fish that should have been allowed to spawn and grow are caught. Government involvement can make matters worse, if subsidies distort signals and prolong overfishing.

A cooperative of fishermen could manage a resource well if it could control its members, if its members could control their organization, and if the cooperative could control entry into the industry-all big ifs. In theory, a private monopoly could also manage a fishing resource efficiently.

More often in practice, government regulates by denying access or enforcing restraint. Restraints that start with and build on existing resource users are more likely to be credible and successful-as seen in New Zealand and Iceland. The Law of the Sea, with 200-mile exclusive economic zones, reflects the idea that privatizing a resource can allow for better management-it "privatizes" ownership to coastal states, suitably linking the resource both to traditional interests and to a natural jurisdiction for enforcement.

While not all fisheries' management problems are confined to national waters, nations can build on the law and negotiate. Many impressive fisheries management schemes would not have been possible without this law. In many other cases, however, management institutions have been only partly successful in limiting entry and stabilizing catches, and as a result there has not been enough reduction in excess capacity and harvesting to bring major benefits.

The challenge is to win the industry's confidence that it can reap the benefits if capacity is reduced and profitability restored. If the commitment to rewarding restraint is not credible, overfishing continues. A good institutional environment must perform many functions. But to credibly and purposefully coordinate actions, three functions stand out-to pick up signals, to balance different interests, and to implement decisions. Fisheries being depleted, toxins poisoning children, corruption weakening emission testing, oppression hobbling indigenous groups, violence against women continuing-all these phenomena can be ignored wilfully or accidentally in a setting that is not receptive to signals. Signals of social and environmental degradation can be based on scientific measurements, or voice and feedback, but they would be effective only if there are constituencies for information and action.

The ability to pick up signals is closely associated with the ability to balance interests. Creating and receiving signals range from the feasibility of detecting a phenomenon in a meaningful way, to the process of aggregating signals and getting the attention of decision makers. Receptivity to signals thus depends on social and political relations.

Both citizens and the air-quality protection agency need good information on pollution, and this information has many uses. Mexico City's IMECA index, published daily in the press, advises people whether to keep their children indoors and avoid exercise. And it obviously informs citizens on how well the agency is doing in improving air quality, strengthening their hand in holding politicians and agencies accountable. But there are also other signals. Voice describes the signals from citizens, firms, and civil society to influence institutions (i.e., through complaints, votes, court proceedings, and the media).

For marginal rural areas, two factors make receptiveness to signals from the fringes more critical now than before. First, outmigration is less of an option than it was when high-income countries were industrializing. Second, other developments, such as mining, happen faster, so that signals about their impact on the community and the environment need to move faster too. Many countries are rearranging the way they govern themselves at the local level. With political decentralization, institutions develop to receive more fine-grained signals-important, since problems and priorities differ from place to place.

But decentralization, proceeding in both rich and poor countries, carries promises as well as risks. It can be hard to get the incentives right, and there may be issues of weak institutional capacity and elite capture at lower levels of government as well.

A protective institution such as an air quality protection agency operates within an authorizing framework that balances interests. It may impose an emission standard or a tax, or strengthen enforcement. This in effect strengthens the rights of beneficiaries to air quality-weakening rights of others, and this is one of the ways changing social priorities can be implemented.

The balancing of interests takes place at many levels: in national legislatures, in court processes, in marketplaces, in individual norms and village interactions, in the seen or unseen processes in corporate boardrooms and branches.

The term "transparency" dominates the current campaign for better governance. There is evidence from rich and poor countries that greater availability of information means better environmental performance. A good working hypothesis-based on studies from industrial and developing countries-is that monitoring pollution information and making it public would help (in part through political channels) to improve air quality in these cities regardless of their level of income. But balance is not maintained by information provision alone. More open and democratic countries presumably give weight to dispersed interests, so there are reasons to expect that they would give more attention to environmental protection.

In China two institutional features play a role in areas such as environmental management, where there has been noticeable progress in the last decade: a systematic approach to complaints, and national policies to make local environmental information available to the public. A complaints-driven system has many qualities, but may be biased toward immediately noticeable phenomena, such as noise. So the combination with objective monitoring data is valuable. Both mechanisms utilize the strengths of a decentralized system and recognize how national institutions can be important for local accountability.

In Europe information that helped establish a common understanding of who suffers from a problem and who contributes to the problem was essential in shifting the balance in favour of reducing trans-boundary pollution, even when negotiation-not authority-did the balancing. For balance and for unbiased signals, it is essential to have supportive institutions in place. Corporations rely on laws and auditors for such traditional goals as protecting workers' pensions and shareholders' as-sets-and they now rely also on civil society for broader corporate social responsibility.

Costa Rica's program to pay for environmental services program is an example where institutions shift the burden of protection, help to balance interests, and ensure better execution. There are many other examples, often integrated with development projects, such as the Global Environmental Facility (GEF) and the Prototype Carbon Fund.

In many countries, laws and protection through the courts are of little value because they are implemented by ineffective or corrupt courts-unless one has connections or money. In others, budget deliberations are not worth much because the budget is not followed. It is not sufficient for society-or a development bank-to make a policy decision. Say society decides that forests should not be converted if the conservation benefits are higher than the conversion benefits. The implementation of this decision can be blunted by developers who press ahead and convert a forest, asserting that the conservation benefits are minor.

The developers count on escaping sanctions-even if the losses turn out to be high-if society is known to lack the incentives, opportunity, or commitment to punish or undo wrongdoing. Good procedures and broad participation can help in the execution of high-level decisions. Procedures requiring ex ante assessments, participation, and public reviews can help.

Routine social and environmental impact assessments, enforced with good quality information and public access to them, can expose consequences before development is irreversible. This can make it clear-to the public, to political leaders, to courts, and to civil society-that the proposed developments do not comply with broader social priorities. The information-and the supportive institutions-function as a commitment device.

In Uzbekistan, as part of the Uzbekistan Water Supply, Sanitation, and Health Project, a social assessment process was undertaken during project preparation. The government initially wanted to ensure that drinking water would have no more than one gram of salinity per liter, although international guidelines allowed higher salinity.

The lower salinity level would have been costly and there is no known evidence that it would be healthier. So with the help of local social scientists, a taste tolerance survey was carried out. It found that salinity levels of up to two grams per liter were socially acceptable. The findings from the exante assessment were accepted by the Uzbek government, and consequent design changes freed up about $15 million dollars. Parts of the savings expanded the project's geographical scope and resulted in additional pilot projects.

Many countries have a gap between the policy decision to teach children with public funds, and the implementation of that decision-to make sure learning is effective. Studies from Argentina, El Salvador, and Nicaragua show that empowering parents can improve the delivery of educational services.

Institutions assist in coordination, but encouraging the emergence of good institutions is itself a co-ordination problem. Three barriers to building and supporting institutions are important:

— Organizing dispersed interests-it takes more for people to come together and build a school than it takes for them to build individual homes. Forging credible commitments-it takes more to commit to helping each other in the future and over a longer period than to helping each other today.

— Promoting greater inclusiveness-it takes more to coordinate for the common good when access to assets and voice are exclusive, unequal, and undemocratic.

A trade regime may display protection even though many may lose from it, since the interests of those benefiting from it are more concentrated and thus more easily organized. Or policies may show an urban bias because the rural population is less vocal and has more difficulty organising itself. Or the civil service may be overstaffed or overpaid because civil servants have a good grip on the policy process. Institutions face challenges in organizing dispersed interests even if the counter-interests are not concentrated.

The collective action to generate institutions for air quality improvements may represent a tremendous challenge even when the interests of polluters are dispersed as well. Protecting air quality requires coordination beyond what the market would accomplish unassisted. Freedom of expression and association, trust, and political accountability provide some institutional machinery to coordinate dispersed interests, both in picking up signals and in giving them balance. Democratic institutions and the popular vote, despite many weaknesses, lower the costs of coordinating dispersed interests.

Society may have an interest in not polluting the water, and it may even prohibit excess discharges in the water. But if the water is already polluted, aggrieved parties may find it difficult or impossible, for political or other reasons, to punish polluters. Such failures to deliver on commitments encourage transgressions by those who are well positioned, and are detrimental to accumulating and protecting assets.

Similarly, a government might want to protect an area from settlement, but it will have difficulty doing so if many settlers can arrive and dig in before the government gets there. A prior action can serve as a commitment device. For example, the United States posted the U.S. army to protect Yel-lowstone National Park against incursions before the park service was established. When assets can be degraded or destroyed without risk of sanction, they are more likely to be depleted, often in a wasteful fashion. But the cultivator settling beyond the agricultural frontier traditionally has not asked for permission.

At low levels of human impact on the environment, anyone can use something that nobody else uses, so this regime makes sense. It allows institutions to advance geographically at a pace with settlements, which is both economic and realistic. But there may also be times when it is important to get ahead of the advancing frontier of settlement, either because there is scarcely any forest, land, or water left, or because selective preservation has become a priority.

In many developing countries the area under agricultural cultivation will stop expanding or may even reverse in the coming 30 to 50 years, so selective preservation now can yield lasting benefits at modest cost. As incomes rise, a time arrives when the forest has value not only to a potential cultivator (or those who depend on it as an ecosystem for their livelihood) but also to citizens far away. If these interests can find a suitable channel for expressing their interests the first-come, first-serve regime can be taken off the table selectively. It is often difficult, however, to fight the tendency to reward early movers. When fish quota systems are put in place, the race for quotas can become a race to the shipyards, or to government offices-races that can be just as wasteful as the race to fish the declining stock.

Preventing waste in environmental matters and natural resources requires confronting those who choose to deplete natural assets with the full social consequences of their actions, whether through informal mechanisms, taxes, or regulations. Similarly, loggers and fishermen must be induced to act as if they faced the consequences of others who might use the forest or the fish. When commitments to protect are not credible, it is highly profitable to move early to manipulate the protective institutions.

While it is well recognized that the quality of institutions affects the management of assets, it is less well recognized that the distribution of assets and voice affects the evolution of institutions in the long term. Poorly distributed assets can affect adversely the quality of institutions and their ability to solve problems. Because of this reciprocal relationship between institutions and the distribution of assets one can get locked into vicious or virtuous circles. These circles are not deterministic, but extra effort is needed to break out of a vicious circle.

Individuals in the community could form an association and negotiate with polluters. Or they could lean on government to do this-in other words, voice and participation are important. A society in which the majority has no voice can lose out big in two ways. First, it can lose because the potential creativity and productivity resting in the majority of the people is ignored or valued only in part. Second, because beneficiaries to communal and natural assets are not heard, the potential of these assets may be wasted, too. Institutions such as the law necessarily involve coercive powers, and one of the potential benefits of broad-based voice-an inclusive democracy-is that it better commits these powers to serving society at large.

Poor people have to accept very costly outlets for their savings, as when they buy gold, are hurt by inflation, or must pay others for safekeeping. And the police and courts, responsible for enforcing the law, often fail to assist or adequately protect poor or disenfranchised groups. Indeed, many institutional development initiatives are geared to making police and judges more attentive to the needs of the poor and disenfranchised. Protecting people-and the emergence of protective institutions for assets When institutions such as the law, and the agencies supporting the law, become more inclusive, more people are given protection, voice, and support.

And when institutions are more inclusive-listening to and supporting more people-a broader range of assets can thrive. The reason is that assets need guardians and spokespersons. Assets therefore may fail to be served if the people who benefit from these assets are not well served by-or represented in-institutions. For private assets, more inclusive institutions facilitate development and asset accumulation as more people feel safe in their homes and find promising outlets for their savings.

For communal and natural assets (roads, water, fish, or forests), more inclusive institutions deepen the support for their provision, so that their quality and quantity can rise. Consider what happened in Cubatão, Brazil, where inclusiveness in the form of democratization and the end of media censorship shifted the balance toward civil society and a cleaner environment.

In many countries, social movements pressing for democratization and environmental improvements have reinforced each other. In a wide range of cases, society relies on guardians, or custodians, to look over something of value. An example is when participatory approaches in projects ask people to speak their mind. The presumption behind this empowering people's voice is not only that people have a right to speak on their own behalf.

It is also that city people, for example, can benefit from hearing from people in more remote areas about what goes on in the forest, about the effects of cutting trees or damming rivers. For people to be functional guardians, they must be well-endowed and feel safe. As an illustration, all societies rely on parents to protect and nurture children. It happens that this protection fails-as when children are sold into slavery or prostitution. This is not because parents are not their guardians-they are-but because of the family's poverty and despair. This need to have well-endowed guardians places broad-based development and poverty reduction at the heart of concerns for environmental and other communal assets.

When people have assets-and thus a stake in the future and in the community-it is also easier to build support for institutions, public goods, and publicly provided goods such as rule of law, watershed management, and schooling. In important ways, high levels of inequality and deprivation can be harmful to efficiency and growth. Economic forces that create differences in income and wealth serve a positive function by creating incentives to allocate resources efficiently. But poverty and inequality can be harmful through other important mechanisms; at the macro level, damage can be done in the political process.

Institutions and government policies are essential for assets to thrive, through the rule of law and macroeconomic stability. An equitable distribution can facilitate the emergence of institutions to negotiate change and thus help adopt and implement good policies-particularly to address externalities and public goods. At the micro level, sharp differences in income and wealth are also costly. Imperfections such as those found in capital markets may allow individual potential to be wasted when individuals are very poor.

Examples are when a talented child goes uneducated, or a worker stays with an employer, or with an asset with low yields only because she cannot finance migration or a job search. Another example is when agricultural potential is wasted because of distorted and highly concentrated landownership and contracting problems. Inequality in land assets has been found to be harmful to growth. Good institutions appear to facilitate long-term growth, and more egalitarian societies appear to have better institutions.

Furthermore, good institutions work in part directly, and in part through schooling and openness. An ambitious quantitative study tested the role of inclusiveness and institutions using a 500-year perspective. For colonized countries, a major break in power structures and institutions happened under colonization. Those that were richer and more densely populated in 1500 are poorer now. This reversal of fortune came about because colonizers in richer, more densely populated areas could force a large supply of labour to work in mines or plantations.

Under these extractive institutions, political power was more concentrated. The lower quality of these institutions for growth reveals itself after 1700, when asset creation became important, rewarding countries that had institutions better suited for savings and investment. Long-lasting harmful effects of institutions that concentrate ownership are also found in a recent study of India.

The British colonization in India lasted for 200 years. Where they implemented a landlord-based revenue system (by implication, with concentrated property rights), yields were higher than in areas where they implemented cultivator-owned rights. In post-independence India, the landlord-based revenue system was abolished, so only the historic traces of the institutions remain. Yields have grown significantly faster in the areas where, historically, cultivators themselves had property rights.

The differences prove to be particularly important from the 1960s onward, as districts with smallholders benefited more from the green revolution, with significantly higher application of fertilizer, high-yield varieties, and irrigation. Districts that historically had smallholder institutions also had higher investments in human capital. The proposition that ownership matters is supported also in other studies. Before 1977 sharecropping contracts in West Bengal, India, generally involved 50 percent output shares to the tenant for the approximately 2 million sharecroppers in the state.

In 1977 a new administration gave high priority to a law giving security of tenure to tenants. The reform increased most tenants' share of output from 50 percent to 75 percent. In the decade after this reform, West Bengal broke through: Annual growth in the production of food grains rose from 0.4 percent to 5.1 percent, while that for all of India rose only from 1.9 percent to 3.1 percent. The tenancy reform program explains about 30 percent of the added growth. Tenancy reform in urban slums in Brazil seem to also have unleashed growth potential and improvements in the urban environment.

In the stages of development, owners of land may benefit from booms more than others. In a similar way, owners may be best positioned to benefit when a community performs better, as when schools and roads are in good repair. For this reason, having narrowly based ownership and many citizens without land or secure tenure can be an impediment to forward-looking and constructive collective action, whether for environmental protection or for other purposes.

Countries with greater inequality have a weaker rule of law (and lower incomes). The key point of this section is that highly unequal access to assets can be punishing to asset creation, preservation, and improvements in well-being if institutions are not rock solid. Groups without assets see themselves as unsupported by property rights, and are thus less supportive of property rights politically. This undermines support for the evolution of institutions that enable growth and sustainable development. A narrowly based elite is often concerned about the risks of more inclusive political empowerment.

One concern is that they might be expropriated. Just as unequal access to assets can be an obstacle to the emergence of good institutions, improperly designed and balanced, redistribution of existing assets can also be harmful to the emergence of good institutions.

If a person without assets is more likely to support expropriation of assets, then a group that has lost through expropriation is also likely to become less respectful of the law and property. It should be clear that it will be easier to improve inclusiveness through access to new types of assets and through the expansion of assets that come from the growth process.

Redistributive measures must be designed and balanced to avoid undermining the emergence of good institutions that enable people and assets to thrive. Important questions are whether institutions are everything, whether policies-in part determined by institutions-have separate and important effects, and finally whether high inequality itself is a major and important obstacle to sound institutions.

East Asian economies was a focus on shared growth, inclusive schooling, and how this served to give political stability and investor confidence.

An important element in political discourse in Western Europe and North America in the 20th century has been "to give everyone a stake in society," supporting policies to strengthen social safety nets, to subsidize general education, and to make home ownership more inclusive. Finally, policies that are frequently pursued-wasteful protectionism, unsustainable macropolicies, a bloated public sector-are best understood as short-term re-distributive games that are costly in the long run.

These games are played at greatest cost in nations with poor institutions, giving them low ability to negotiate and to commit to mutually beneficial change. Institutional reform happens when the actors take advantage of opportunities for change, and use instruments of change at their disposal. The institutions that mediate social interaction must foster both stability and change. A measure of stability and predictability in the rules governing society is necessary for the people to have confidence to work together, to challenge each other to improve their communities, and to invest in their future.

A vibrant civil society and such institutions as a democratic legislature can provide for dynamism-including that in rulemaking. In a society founded on broader consensus and certain ethical principles, these institutional are simultaneously given force and anchored to give predictability and confidence. A seeming paradox is that democracies-despite their frequent leadership changes-can better commit to the longer run, and do the right things. When characterized as a stable democracy, countries benefit from predictable successions, as if the democratic institutions themselves have taken on the role of an owner with a long-range perspective.

However, studies show that young democratic states have problems similar to those in autocratic states, in terms of protecting property rights to allow assets to thrive, at least until democratic institutions have taken root. This is a challenge for fledgling democracies. Many East Asian countries have done well in the last 30 years in stability and asset creation, but not necessarily through democratic institutions. What these countries had in common, however, was broad access to land or other characteristic that made them emphasize shared growth, such as through broad-based rural development and broad provision of health and education. Some institutions-such as constitutions-are designed to make change exceptionally difficult.

Constitutional changes typically require a much higher degree of voter consensus than do lesser legislative reforms. And a two-chamber legislature requires that coalitions be built in alternative ways for changes to pass. Other institutional means of providing stability and commitment are so-called checks and balances, that is, independent veto players. High court justices appointed for life, and systems with separation of power among branches of government can ensure that radical departures from the norm are not made in haste.

Institutions may also be designed to facilitate change in one direction and not in another. In many Latin American countries a national park may be created with a simple presidential decree, but dismantling a park requires the approval of both the president and the legislature. Some countries have anchored their commitment to the environment in the constitution by linking the environment to the rights of their citizens. Others, such as India and Pakistan, have supreme court decisions serving the same purpose.

All these institutional mechanisms depends on whether they rest in shared values, so that relevant actors abide by them. In general, multiple review procedures and systems of checks and balances work only when political and economic power are not too concentrated. Improvements in social conditions and in the institutions that shape them often seem unbearably slow.

In Latin America, perceived crises in the legitimacy of governing institutions have inspired substantial democratizing reforms that give greater voice and power to local communities. On a more routine basis, opportunities appear as a result of elections, changes in agency leadership, or discretionary decisions by national leaders. And opportunities for reform arise with changing public preferences, and with changes such as education, urbanization, technological change, and income growth. For example, new generations of individuals raised in conditions of material prosperity and stability tend to place greater emphasis on freedom of expression and quality of life.

The demands that societies place on their institutions also change as a result of observing other societies. In recent years the transnational social movement for indigenous rights, the sweep of independence movements across Eastern Europe, and the multinational campaign for transparency in governance show that new social demands can spread rapidly across borders.

Unaccountable power structures are often vulnerable-and thus potentially responsive-to the exposure of information, so having more information available can be a catalyst for change used by reformers, entrepreneurs, and civil society. A broad set of studies shows the power of information disclosure in environmental performance, in both developed and developing countries. A growing number of firms and NGOs provide investors and markets with information on the social and environmental performance of companies-information that facilitates decisions by investors and customers, and catalyses changes in the behaviour of firms.

Enhancing the voice of stakeholders can dramatically shift the balance of forces that favours institutional reforms. Cultural translators can bring new ideas and ways of working together that increases the self confidence and voice of groups previously excluded from participating in or authorizing institutional change.

In Cubatao, Brazil, the state environmental protection agency was able to reduce pollution by joining forces with a vocal citizens' movement during Brazil's transition to democracy. Stakeholder forums can facilitate conflict resolution and the sharing of ideas, and the consensus building associated with international environmental treaties has allowed steady progress on issues ranging from wetlands conservation to preventing oil pollution at sea. Some changes come about easily, some only with other structural changes-such as changes in technology, endowments, and values.

Changes in labour market conditions, combined with public and government action, have contributed to phase out slavery and elevate women's status. And women with more voice, clout, and knowledge have delivered fertility declines and better-educated children.

Economic growth, better transport, and the successful growth of cities give poor people a broader choice of employers, neighbourhoods, and service providers. Today, perhaps the most significant structural change is the information revolution, with the potential not only to increase knowledge and the use of knowledge but also to improve accountability.

Dynamic development is sustainable when it is forward looking and responsible. Therefore it must be assessed not only by such indicators as poverty, natural resources, forest coverage, and ocean temperatures but also by the institutional environment that helps this information emerge, gives it weight, and ensures that it is acted on.

Institutional Barriers

In the modern period, few variables are as strongly correlated with a country's level of development as its level of urbanization. Among the colonies of the Americas, only the US and Canada provided the "social infrastructure"-the collection of laws, institutions, and government policies-that made such investment and participation possible. The most systematic studies to date of paths of institutional development have focused on the New World economies. In the New World colonies, factor endowments, broadly defined, determined the level of inequality of income, wealth, and human capital at the beginning of colonization.

In turn, the level of inequality had a great influence on the evolution of institutions. In highly unequal colonies, institutions emerged that blocked effective access by a broad cross-section of the population to opportunities for economic and social advancement. This inhibited the accumulation of capital, the spread of entrepreneurship, and the creation of mass market-factors that are viewed as important in industrial development. The New World colonies can be put into three broad classes in terms of their factor endowments at the outset of colonization.

In many parts of Spanish America, e.g. Peru and Mexico, there were valuable minerals and large concentrations of populations of Native American descent that survived contact with the colonizers. Spain adopted a policy throughout Spanish America of distributing vast grants of land, including claims on native labour in the vicinity. Spain also limited the immigration of European settlers, which contributed to the persistence of elites and the maintenance of vast landholdings even when the production activities (as in Argentina) were not characterized by economies of scale.

Climate and soils well-suited to the production of sugar and other highly valued crops cultivated with slave labour. In the Caribbean, factor endowments were suitable for large plantations based on slave labor. The large slave population and the very unequal ownership of land among free men made the distribution of wealth, income, and human capital extremely unequal, both among the population as a whole and among the population of free men.

In North America, there was no preexisting dense population of Native American descent. Except for the South, the colonies had soils and climates that did not give them a comparative advantage in the production of crops using slave labor. The result was that development relied on labourers of European descent. Such labor was scarce, as the colonies of North America were not viewed as particularly attractive places to settle. Given the technology of the time, grains could be profitably produced on very small farms. The approach to colonization of the British on the North American mainland was not very different from its approach in the Caribbean. Nor was it very different from that adopted by the Portuguese in Brazil or the Spanish in Spanish America.

In general, at the outset of colonization, the British in North America established large estates and oligarchic governing bodies. But because of the labor scarcity, the situation of large estates and oligarchy did not take hold. In the southern colonies of what became the United States, conditions were somewhat different. Rice, tobacco, and cotton exhibited scale economies, but even here the slave population and the degree of inequality were much smaller than in Latin America.

Although all the economies established in the Americas had abundant land relative to labor, their climates, soils, and minerals differed in ways that meant that most were characterized at the outset by extreme inequality, whereas the colonies that were to become the US and Canada did not. There are systematic patterns across New World economies whereby institutions in societies with greater initial inequality tended to evolve in such a way as to restrict access to opportunities, favouring elite groups.

The factor endowment and the degree of inequality may influence the directions in which institutions evolve, but these institutions in turn affect the evolution of the factor endowment and of the distributions of wealth, human capital, and political power.

Most societies in the Americas were democracies by the middle of the 19th century, but there were sharp differences among countries in the breadth of the respective populations having effective access to the vote. Until early in the 19th century, all countries , including the US, limited the right to vote to white men. This pattern also occurs within countries, as seen for example in the case of the US South, which used a wide variety of means, including literacy tests and few and relatively inaccessible voting locations, in order to limit the access of the poor to the ballot.

In the Western states of the US, labor scarcity played a role in the evolution of rules regarding suffrage, as the pioneers in extending the franchise were states that were competing for migrants. Most Latin American countries did not publicly provide primary schooling on a scale sufficient so serve the general population until the 20 th century.

In contrast, public primary schooling funded at the local level was very widespread in the early 19 th century in the US and Canada. An alternative explanation focuses on the role of inequality. In a highly unequal society where the vote is limited to the wealthy, the voting elite faces some trade-offs.

On the one hand, providing mass education will raise the productive potential of the poor majority, which promotes growth and may also create positive spillovers for the wealthy. On the other hand, the cost of that education will be borne disproportionately by the wealthy. Further, the newly educated poor may agitate for the right to vote, which will threaten the political power of the elite and their control over government policy.

Among the New World economies, empirical work shows that economies with greater inequality in political power (as reflected in the proportion of the population who vote) have a lower fraction of the population enrolled in school, controlling for per capita income and time and region. Among all economies today, those with greater income inequality are likely to have lower expenditures on schooling, particularly on primary schooling, and are likely to be less democratic.

In most of Spanish America, land and labor policies and restrictions on immigration led to large landholdings and thus maintained great inequality. In contrast, the US and Canada maintained open immigration and offered small units of land at low prices to households who farmed the land.

In general, households without land or other wealth to pledge as collateral against loans have very limited access to credit for investment purposes. Consider an economy where every individual has some talent that he can exploit more effectively as a proprietor than as a wage labourer, but that setting up a business is not possible for a poor individual because he cannot obtain credit to finance his fixed costs.

When the ownership of assets is highly concentrated, only a few will be able to employ labor, while the many without assets must supply labor. Thus wage rates will be lower than in an economy with a less concentrated distribution of wealth. With low wages, incomes of those without assets will be low, and so their bequests will also be low: the initially highly skewed distribution of wealth, which blocks opportunities of the poor to participate in commercial activities as proprietors, may be self-perpetuating.

Economies with the same aggregate endowments and the same technology but different initial distributions of endowments may follow very different paths of development. Economists traditionally believed that the only institutions that mattered for economic development were those that defined property rights. The distribution of those rights across the population did not matter for aggregate growth. Related to this view, there was a tendency to adopt the functionalist position that efficient institutions would "naturally" emerge over time. The most famous expression of these views is the Coase theorem, which holds that regardless of the initial allocation of property rights, individuals will make agreements that lead to efficient outcomes.

Capitalism under the rule of law can be viewed as a game in which the same rules apply to all, property rights are secure and enforced, and areas of ambiguity are resolved through predictable unbiased processes. No country has ever attained this ideal; its realization is always a matter of degree

In the eastern region of Indian state, landholding sizes are heterogeneous. The wealthier members of the sugar cooperatives, who wield disproportionate control rights, set the prices for sugar supplied by members at a low level and divert the retained earnings to their own benefit. With low sugar prices, the farmers have low incentives to improve productivity.

In contrast, in the western region of the state, landholdings are in general uniformly small. The cooperatives set the prices for sugar supplied by members near the world price. With high sugar prices, the farmers have high incentives to improve productivity. Growth is higher and capacity is more fully exploited.

The agricultural tenancy reform implemented by the Indian state of West Bengal after 1977 permits a quite different test of the consequences of the distribution of property rights on efficiency-a test of the effects of a redistribution of rights to output in favour of tenants. Before 1977, sharecropping contracts in West Bengal generally involved 50 percent output shares to the tenant for the approximately 2 million sharecroppers in the state. In 1977, a new administration was elected. One of its highest priorities was to enforce a long-dormant tenancy law that stipulated that (a) the tenant would be given the choice of registration with the government, and (b) those tenants who have registered themselves cannot be evicted form the land as long as they pay the landlord a minimum legally stipulated share-rent.

The result of the reform was to increase many tenants' output share from 50 percent to 75 percent, and to give them permanent inheritable tenure of the land they sharecropped. In this period, West Bengal changed from one of the worst to one of the best states of India in terms of agricultural output.

The landlord has all the bargaining power in the relationship. This is a critical assumption, but it is not unrealistic in a setting such as exists in some villages in India today or historically existed in some Latin American countries. Suppose that the landlord can observe the tenant's assets, but cannot perfectly observe his effort. The tenant is so poor that he cannot absorb losses. In order to provide him with incentives to work, the landlord, in his own self-interest, provides him a labor contract whose terms are more generous than those of any alternative employment of the tenant.

The tenant farmer receives his income, out of which he can save. Now consider the second period. The tenant farmer again negotiates a contract with the landlord. If he has saved some of his income from the preceding period, the landlord (seeing that the tenant has some savings and therefore can absorb more losses) provides a less generous contract than the tenant earned in the preceding period. As a result, savings face an implicit tax:

For limited degrees of wealth accumulation by the tenant, it is the landlord, not the tenant, who benefits. Thus, the tenant has no incentive to save, and his escape route from poverty may be blocked, no matter how long the term of his productive life. But at higher levels of wealth, tenants are able to absorb losses, the tenancy contract provides no surplus, and the incentive to maintain wealth is high. In the long run, the wealth distribution may become polarized into two classes, with no middle class nor any interclass mobility.

A high initial level of inequality of wealth and a dense population of labor with little human capital, institutional barriers can emerge that restrict the opportunities of the poor for economic advancement, and that protect the elite. Those barriers can sometimes be understood as the outcome of political forces, and sometimes of market forces. In turn, these institutions affect the evolution of factor endowments and the distribution of wealth in the economy. Under some conditions, they will reproduce the conditions that gave rise to them. In this way, institutional arrangements that restrict opportunities for growth may persist.

Patent institutions provide inventors temporary property rights to their inventions and facilitate trade in the rights to inventions. Thus they make possible a market for new technological knowledge. Social interactions also play an important role. Theoretical and empirical work shows that sorting in the marriage market makes household inequality greater than individual inequality. Household inequality is the most relevant measure of inequality in the determination of investments in children's' education.

Process of obtaining information was sufficiently difficult that patent agents were generally used as intermediaries. They were few in number and were able to keep prices for their services high. Indirect econometric evidence suggests that the relationships are causal. Relative to the US patent system, the British system more tightly restricted access to intellectual property rights in ways that had the consequence of limiting the class of inventors to those with substantial wealth, political connections, or technical knowledge.

Broad-based property rights to technological information in the US spurred an enormous increase in inventive activity in all sectors, in virtually all sub-regions, across a broad spectrum of the population. The overall pattern illustrated in the table holds both for a sample of all inventors, and for the sample of "great" inventors, defined as those whom histories of technology credit with at least one important invention.

Evidence also points to important interaction effects with public infrastructure. The building of canals, such as the Erie Canal, caused patenting activity in the vicinity to boom, as detailed maps of patent awards illustrate. The finding suggests a link between patenting and access to extensive markets, because waterways were the only means of low-cost transportation over long hauls until the 1830s.

Rural counties especially realized large and rapid increases in patents per capita as they gained access to waterways. For them as for the small inventors, the expansion of markets during early American industrialization induced a broad segment of the population increasingly to commit resources to inventive activity which in turn raised the rate of technological change. Far from being exogenous, the activity of the great inventors was influenced by much the same market-related forces as invention by ordinary patentees.

The comparison of US and UK patent institutions implies that the details of patent institutions are very important in influencing the breadth of the population that has effective access to intellectual property rights. In each country, the same rules applied to everyone. In that sense, each legal regime looked symmetric. But what looks symmetric is not symmetric because the UK system imposed what would appear to be needlessly high transaction costs, and their burden was much greater for the poor and the politically unconnected than for the rich.

South Africa has one of the highest crime rates in the world, but the crime rate is highly variable across police districts. A study examines the relationship between home-ownership rates and crime, controlling for a wide number of relevant variables: income level and inequality, education, race, occupation, residential segregation, female-headed households, and urban areas. Controlling for these variables, the study finds that crime is lower in districts where the home-ownership rate is higher. This pattern holds for crimes against property, such as burglary, as well as violent crimes, such as murder. An increase of one standard deviation in the home-ownership rate is associated with declines of 14 % in the burglary rate and 18% in the murder rate. This may be because of omitted variables, but the findings are consistent with the view that home-ownership, perhaps by promoting vigilance or community solidarity, has a causal effect on reductions in social disorder.

Studies in developing countries also show that owners participate more than renters in community upgrading but we are not aware of studies that control for differences between homeowners and renters. Thus it is not possible to assess whether individuals participate more because they are homeowners, or whether the characteristics of individuals that lead them to choose to be homeowners also lead them to participate more in community upgrading. Titling is one aspect of security of property rights, but in many cases titling may have only a small effect on tenure security because there exist informal mechanisms that create tenure security.

The relationship between residents' civic participation, incomes, and the choices that lead to a given structure of property rights in a community. They consider an economy with three features:

(1) residents have an incentive to expend effort to improve their communities only if they are homeowners,

(2) capital market imperfections make it more costly for low-income households than high-income households to own, and

(3) for an individual household, the marginal returns to effort are higher when neighbours also engage in community improvement efforts.

Households that are identical in all respects except income self-organize into "good" and "bad" communities. This simple model shows that market forces and social interactions may create a situation where poor households are isolated in communities where no resident has a large enough stake in the community to participate in upgrading or to form the kinds of neighbourhood associations that can solve problems of collective action and give the community political voice.

3

Virtual Resources

Certain types of natural resources such as oil and minerals have a tendency to lead to production and revenue patterns that are concentrated, while revenue flows from other types of resources such as agriculture are more diffused throughout the economy.

Economies with point-source or concentrated resources such as oil and diamonds can be expected to have a higher propensity to engage in rent-seeking and other unproductive activities when compared to economies with diffuse resources such as fertile soils, grazing land, and fisheries.

There is a greater chance of a vicious cycle of mismanagement, rent-seeking and even conflict in countries with resources that can be more easily expropriated. Economies that are coffee/cocoa based may also be regarded as point-sourced, as these commodities tend to be produced in plantations with concentrated ownership.

A country is point-sourced or diffuse based upon its principal export commodity, not ownership patterns or the share of the natural resource sector in national income. Thus, it is anomalous that Zimbabwe is considered diffuse because its principal export is tobacco, a diffuse crop, whereas land ownership is more like a point-sourced economy. The same may be said for the Philippines.

If natural resource endowment, particularly of the point-sourced variety, retards economic growth and development, two questions immediately spring to mind. First, what are the channels between this type of riches and underdevelopment? Secondly, is this resource curse inevitable, and what can be done to counteract it? Inappropriate or bad policies, in turn, are the products of poor institutions. The avoidance of the resource curse, therefore, requires fashioning good institutions of governance that produce appropriate policies.

In an open economy a substantial current account surplus appears, leading to currency appreciation under a regime of flexible exchange rates. This renders existing exports even more uncompetitive in world markets. A shift in domestic output from tradeables to non-tradeables takes place. Temporary resource booms cause hysteresis, a permanent loss of competitiveness. For developing countries, this means that their future potential for exporting manufactured goods and diversifying the production base is stunted.

Human capital accumulation, in the form of an externality, takes place as a result of traded/ manufacturing production only. In contrast, one of the earliest treatments of human capital accumulation is in Shell. Knowledge is a public good, the accumulation of which benefits all sectors in the economy through cost savings. A resource boom unambiguously expands the non-tradeable sector, while at the same time shrinking the traded sector. If it is the expanding sector that uses these intermediate inputs, it may contribute to a successful big push. If the opposite is the case, and it is the traded sector which uses the intermediate inputs, big-pushes are less likely. Unless expectations about the future are optimistic, even the most propitious circumstances may not trigger accelerated industrialization or the big-push.

Capital is a specific factor in manufacturing, whereas land is specific to the resource sector. All sectors require labour input. In these circumstances a resource boom will induce a lower optimal supply of the publicly financed input, as the resource sector obtains a proportionately smaller benefit from this input. Consequently, over the course of time, both sectors will be less productive, akin to a loss in absolute advantage in international trade. The expansion of international trade will also make countries with greater capital endowments gain absolute advantage in all sectors, as exports of manufacturing increase, inducing greater provision of the cost reducing public good.

If an additional, non-traded and publicly supported consumption sector is introduced, similar to the functioning of state owned enterprises, resource booms will retard competitiveness in both the other sectors even further in the presence of a strong societal or ruling class preference for this good. The reason is that the reduction of the supply of the publicly financed productive input is greater after a resource boom in the presence of a strong preference for a publicly supported non-traded consumption good.

Manufacturing employs labour and sector specific capital. Production in the resource sector utilises labour and a sector specific land input. Land utilised in the natural resource based production sector is not just exogenous, but the land frontier (and the output of the resource sector) can be extended by the application of capital input. Globalisation in the 19 th century lead to a rise in the demand and price of primary goods produced both in the tropics and the regions of recent settlement. This not only raises the rental rate on land used in primary goods production, but also extends the land frontier. It also raises the demand for labour in the new world and in tropical regions, and leads to immigration into the regions of recent settlement, and a movement away from subsistence farming to cash crops or mining in the tropical regions.

Manufacturing, too, expands in Europe and in the regions of recent settlement. The mechanism of expansion for manufacturing is a decline in the real rate of interest. This raises the capital intensity in manufacturing, as well as the real wage rate. In tropical areas there is also an increase in the real wage in the primary goods sector, above subsistence sector levels. This wage premium is necessary to finance additional land clearance, allowing the resource sector to expand in the absence of foreign investment. Even with foreign investment, a wage premium may be necessary, reflecting higher productivity.

In plantation type economies (point-sourced), for example in Latin America, landlords capture the wage premium, whereas in peasant, owner-occupied diffuse type societies, the extra rent accrues to peasant entrepreneurs. The regions of recent settlement will be on their way to a bliss point of high per-capita incomes that we nowadays associate them with, because of the backward and forward links to manufacturing, competitive industrialisation in contemporary parlance.

The tropical regions may never industrialise, stagnating instead into a staple trap (a fixed reliance on a few commodity exports). This outcome is more likely in point-sourced economies. By contrast, it is the diffuse economies, such as in North-East Asia, where prospects of industrialising are more promising. These economies move up the manufacturing product cycle. Moreover, peasant societies are also more likely to support publicly financed infrastructure and human capital formation, compared to countries dominated by a small elite interested in siphoning off resource rents. Also, when point-sourced economies experiment with industrialisation, it is usually capital intensive and dependent on public subsidy.

Latin America, when compared to East Asia, exhibits a pattern of growth associated with natural-resource abundance. The former comprise point-sourced economies and the latter diffuse. Latin American long-run growth rates will be less impressive than in East Asia. The traded good is labour intensive, whereas the non-traded good utilises an imported intermediate input making it the capital- intensive sector. If a resource boom, which can occur for a variety of reasons, takes place it will not automatically cause the traded sector to contract and the non-traded sector to expand.

Under East Asian conditions of a high propensity to consume the domestically produced traded good, that sector could expand. With regard to policies to tackle the effects of resource booms, devaluation may be a policy initiated to avoid the adverse effects of resource booms.

There is the possibility of contractionary devaluation, particularly for the non-traded sector. This likelihood is strongly associated with Latin American or point-sourced economies.

When devaluation is expansionary, it is so because the non-traded sector is less important to domestic consumers and there is a sharp reduction in imported consumer goods. These are more likely in the East Asian or diffuse resource-rich case. Policies to tax non-traded goods consumption will be akin to an industrial policy favouring the production of traded goods. Such a policy would be initiated because of the view that labour-intensive manufactured traded goods are superior to non-traded goods production, the latter incorporating manufacturing 'dinosaurs' from the past. This policy is most likely to succeed when non-traded goods are quite price-elastic in demand, and the propensity to consume them out of income is small. Arguably, these are features of the more successful East Asian economies with their diffuse production structure.

Country size would also be an important consideration. Without a critical mass of consumers geared to the domestic consumption of labour- intensive traded manufactures, industrial policies of this type cannot succeed. Growth failure represents the malfunctioning of the institutional superstructure. The equilibrium level of the capital stock in the steady state implicitly defines growth rates, and a fall in the equilibrium capital stock implies a decline in the growth rate. Growth collapses are associated with periods of declining capital accumulation.

The rule of law tends to be the most important institutional characteristic in explaining human development, particularly in Latin America. If East Asia is taken alone, the quality of the civil service is the most important factor. Furthermore, in Latin America the quality of the bureaucracy and the rule of law are often substitutes into the good-governance production function; whereas in East Asia it is strong civil society and the rule of law that are often the substitutable inputs.

Resource booms and windfalls increase the appetite for transfers, within these powerful coalitions, by a factor that is more than proportionate to the size of the boom. These groups become greedier, and demand an even larger share of national income. Natural resource rents can make corruption, predation and rent-seeking a more attractive option. This incentive is greater the weaker the environment of law and contract enforcement following societal upheavals, such as in Russia.

Furthermore, societies in transition, or in the phase of post-conflict reconstruction, often experience a rise in criminal predation, particularly if there is a lot to loot, as in point-sourced natural resource abundant countries. Moreover, the rise in criminality encourages a parallel protection industry offering security services.

In other words, it is difficult to design mechanisms that allow parties to commit to policies, including optimal policies, as they have an incentive to renege on pre-announced commitments. This reduces the credibility of policy pronouncements and commitments. It should be noted that transaction-cost politics constitutes an additional problem, over and above the standard difficulties associated with agency such as adverse selection and moral hazard.

Transaction-cost politics is a general problem, not peculiar to developing countries or resource rich economies. The amelioration of the problems associated with it in a badly functioning institutional environment requires mechanism design and constitutional change. In other words, a sea change is necessary, deep interventions that alter the rules of the game and the processes by which policies are actually formulated.

Over the last 5 years or so, economists have started paying greater attention to internal conflict. This new research has very much arisen out of a pressing need to understand continued development failure, especially in Sub-Saharan Africa. In this new literature, a distinction is often made between grievance, a motivation based on a sense of injustice in the way a social group is treated, often with a strong historical dimension; and greed, an acquisitive desire similar to crime, albeit on a much larger scale.

Extreme poverty and poor social conditions also facilitate conflict by providing more readily available combatants. Competing groups are formed when collective action problems can be overcome. Many of today's civil wars have an ethnic or nationalist dimension and ethnicity, whether based on language, religion or other distinctions, is often a superior basis for collective action in contemporary conflicts in poorer countries than other social divisions such as class. In coalescing groups, therefore, present-day and historical grievances play a crucial part. This is all the more possible when there are inequalities across a small number of clearly identifiable groups-horizontal inequalities.

More often than not, these take the form of high asset inequality, discriminatory public spending across groups and unequal access to the benefits of state patronage. Furthermore, state failure in providing security and a minimal level of public goods often force individuals to rely on kinship ties for support and security, this ethnic capital thereby becoming highly important to individuals of greed as a motive for conflict has mainly arisen in the context of natural resource endowments, an abundance of which-at least so far as valuable minerals are concerned-appears to increase the risk of a country falling into serious conflict.

But poverty also plays a part in the model by lowering the cost of participation in the civil war. But for all of these forces to take the form of large-scale violence there must be other factors at work, specifically a weakening of the social contract. Therefore, while rents from capturable point resources do constitute a sizeable 'prize', violent conflict is unlikely to take hold if a country has a framework of widely-agreed rules, both formal and informal, that govern the allocation of resources, including point-resource rents, and the peaceful settlement of grievances.

Such a viable social contract can be sufficient to restrain, if not eliminate, opportunistic behaviour such as large-scale theft of resource rents, and the violent expression of grievance. The viability of this social contract, in turn, depends on the existence and functioning of good institutions.

Clearly there is now a consensus of opinion that the framework of governance, including respect for property rights, contract and law enforcement, the rule of law and administrative capacity, matter a great deal if a country is to be successful in its quest for growth and development. The current literature points out to three empirical sources of institutional determination, all three of which relate to natural resource endowment.

Countries should not be classified to be in the point-sourced or another category on the basis of their exports alone, but also on the GDP share of these important sectors. Indeed all developing countries can be found to have grown at a similar pace in the two decades before 1973. Thus, the type of resource endowment seems to matter in post first oil shock period. The problems of the predatory state may have become worse following the slump in commodity prices that has steadily occurred since the early 1980s, as there are fewer resources to go around exacerbating the nature of predation.

Many developing nations have had a collective experience prior to, and after, colonialisation that must have also shaped institutions. In East Asia, South Asia, the Middle East and North African regions of the developing world, well functioning institutions of good governance existed well before the advent of colonialisation, and European colonial powers merely adapted preexisting administrative institutions.

Commodity endowments of the point-source variety tend to depress the middle-class share of income in favour of elites, as in Latin America. The idea being that these elites, in turn use their power, identical with the forces of the state, to coerce and extract rents. When different groups compete with another for these rents, the rent-seeking contest analysed in the previous section becomes applicable, and can lead to even more perverse and wasteful outcomes than when elites collude.

But not all point-sourced economies go through, or stay with, persistent coercive behaviour by elites. The obvious exceptions are in the diffuse economies of North-East Asia. But, in the point-sourced economies of South-East Asia such as Malaysia, or to an extent in Indonesia, the elites opted for redistribution. The difference here was that the danger of an ideological upheaval in terms of communism was much more real than in Latin America, and the ruling class wanted to avoid ethnic strife.

Botswana, too, avoided ethnic fractionalisation, unlike in the rest of Africa. In Latin America ethnic minorities have usually been suppressed, and processes of conciliation avoided. It has to be said that the poor institutions associated with extractive industries or colonial coercive policies are not immutable, they can be altered unlike the fatalistic implications of some of the econometric models outlined above.

Resource booms may be engineered to contribute towards competitive and sustainable industrialisation, avoiding the staple trap. The idea is to be adaptable to economic change. Most importantly, an uncompetitive domestic non-traded manufacturing sector, similar to the many state owned enterprises in developing countries, needs to be avoided. It is important to utilise resource rents to develop human capital via education, and infrastructure as in transport and telecommunication networks. This ought to be continued even when the resource boom peters out, via careful prioritisation of public expenditure.

A well developed human capital and infrastructural stock is what allows a country to respond positively to change in the economic environment. Well functioning fiscal institutions are crucial because resource booms are about revenues. These institutions are important if revenue is to be collected, leave alone spent wisely. Many point-resource rich countries, who are also in a state of civil war, such as Angola and Afghanistan, have allowed their fiscal institutions, especially related to revenue collection to degenerate greatly.

In Indonesia and Malaysia policies of competitive industrialisation were also followed, as opposed to the unsustainable state owned enterprises, common to many countries. Conflict avoidance via a sustainable social contract should be a major policy goal in point-sourced economies. The literature on contemporary civil wars lists two major reasons underlying contemporary conflict. The first is to do with intra-group or horizontal inequality. The second is concerned with acquisitiveness, the desire to capture resource rents. Both these features are more likely in point-source economies. But ultimately, open conflict reflects state failure and the breakdown of the mechanisms of conflict prevention, the social contract. Malaysia and Botswana provide recent examples of multi-ethnic states that avoided not only economic mismanagement, but also natural resource contests degenerating into civil war.

Consequently, strategies of poverty reduction by themselves are insufficient, attention has to be focused on lessening inequality, in the sense of the middle-class share, if growth is to take place and be truly pro-poor. Also, the reduction of group or horizontal inequalities is important in outright conflict prevention in low- income countries endowed with point-sourced resources and weak institutions. Each nation is a policy experiment. Whether it adopts good policies or not depends on its institutions, and hence the importance of the institutional framework that determines policies.

Institutional malfunctioning, and the absence of institutional restraint on greed can have severe growth retardation effects in resource rich countries. Lasting institutional change cannot be brought about without societal ownership; in other words it cannot be effected from without and must come from within the nation. Changes to affect institutions can take the form of deep or shallow interventions.

Sustainable Development in Mountains

Mountains provide a substantial portion of the world's timber and minerals, and mountain peoples' environmental services are critical to the sustainability of their lowland "plains." They shelter over half of the world's biodiversity and nurture rich and varied cultures that have much to teach the rest of the world about sustainability and natural resource management.

Mountain people live in remote, rugged, and hard to reach environments. Their dispersed settlement patterns make it difficult for them to form a political constituency; so they frequently lack access and voice in political systems. They are all too often ignored by governments, by traditional development agencies, and by most of the non-profit community. If we do not solve the challenges of conservation, poverty reduction, and sustainable development in mountains, there is little hope of solving the challenges of sustainability anywhere.

Mountainous areas are generally characterized by their altitude, steepness of slope, accentuated relief, and prominence with respect to surrounding landscapes. They exist in a wide range of latitudes, from the polar regions to the temperate, sub-tropical, and tropical zones. Mountains constitute important and distinct regions in many countries and regions throughout the world.

Table; 1 Mountain Countries

1 (%mtns)	Countries and territories of the world and their mountain areas	2 (no.)	3 (%)
0-25	(Countries in this category are not listed)	101	4.4
25-50	Bolivia, Bulgaria, Colombia, Croatia, Cyprus, Czech Republic, Dominican Republic, Ecuador, Ethiopia, Fiji, French Polynesia, French Southern & Antarctic Lands, Greenland, Guatemala, Indonesia, Israel, Jamaica, Jan Mayen, Madagascar, Malawi, Malaysia, Mauritius, Mexico, Mongolia, Myanmar (Burma), Pakistan, Panama, Papua New Guinea, Peru, Philippines, Portugal, Puerto Rico, Romania, Russia, Sao Tome and Principe, Solomon Islands, South Africa, South Korea, Spain, St. Vincent and the Grenadines, Thailand, United States, Venezuela, Vietnam, Yemen, Zimbabwe	46	9.5
50-75	Afghanistan, Albania, Antarctica, Austria, Azerbaijan, Burundi, Cape Verde, Chile, China, Comoros, Costa Rica, Djibouti, El Salvador, Eritrea, Greece, Haiti, Honduras, Iceland, Iran, Italy, Japan, Laos, Morocco, New Caledonia, New Zealand, North Korea, Norway, Serbia, Slovakia, Slovenia, South Georgia and the South Sandwich Islands, Svalbard, Swaziland, Taiwan, Turkey, Vanuatu, West Bank, Western Samoa	38	9.6
75-100	Andorra, Armenia, Bhutan, Bosnia and Herzegovina, Georgia, Kyrgyzstan, Lebanon, Lesotho, Macedonia, Montenegro, Nepal, Reunion, Rwanda, Switzerland, Tajikistan	15	0.5
0-100	Total countries analysed	200	24

Legend:

1 Percentage of mountain areas within a country (four categories)

2 Number of countries analysed per category

3 Percentage of total land area in category (total global land and ice area is 144 million, of which about 36 million km 2 are mountainous)

Source: Adapted from Kapos et al., 2000

The vast stores of fresh water in the polar ice caps, which could become a significant factor in sea level rise with global warming, it is estimated that more than half of the world's fresh water comes from mountains. Indeed, all of the major, and many of the world's minor rivers originate in mountains. Well over two billion people in India, Pakistan, Nepal, China and the Indochinese peninsula depend on flows of the great rivers originating in the Himalaya.

The city of Los Angeles could not exist without water that has its source in the Rocky Mountains; the people of Australia and Japan depend on water from mountains, and the Arabian Desert Peninsula is reliant on rivers that flow from the Jeman and Oman mountain chains. The Nile, which provides 95% of Egypt's fresh water, has multiple tributaries in mountains, from Ethiopia to the Rhuengheri range of Rwanda. Almost all of South America relies on water that has its source in the Andes. Based on the definition of mountains outlined above, some 2.4 billion people live in mountain areas. This is 42% of the world's total population. With the exception of the of the Upland Plateaus, which tend to be quite isolated and arid, population densities in Mountain areas tend to be higher than in lowlands. This is true even for zones classified as high and very high mountains. This concentration of people in such rugged areas has implications for their subsistence and the sustainability of mountain production systems.

Agricultural potential in mountains is limited by the small size of arable plots, climatic variability, and more difficult growing conditions, typically including shorter growing seasons due to altitude. These areas are unlikely to be as productive of basic food crops as lowland areas, contributing to higher levels of poverty in mountains.

Much of the population in high altitude zones is probably living in valleys. They are nonetheless affected by the isolation of these high relief areas and highly dependent on their immediate mountain surroundings.. Indicators of cultural factors (frequency of different languages, different ethnic groups, different religions, numbers of sacred sites)

Following are the threats to mountain and their people:

— Rates of deforestation and cumulative loss.

— Rates of additional siltation downstream and negative effects.

— Impacts of increased water flow variability (flooding) where it occurs. Amounts of downstream pollution from mining and other such activities and their impacts.

— Estimated rates of loss of biodiversity.

— Estimated rates of deterioration of cultural sites and amenity values.

In the mountains, more than half of wars and armed conflicts have taken place. Many of these have been determined to be consequences of environmental changes linked to natural resource degradation, poverty, and related social and cultural strife. Ecological vulnerability inherent in mountain areas plays a major role; and this is exacerbated by competition for mineral reserves, surface resources, and water. Moreover, the inaccessibility of mountains makes them particularly prone to use by guerrillas and terrorist groups, such as the Shining Path in Peru, the Maoists in Nepal, and more recently, the Al Quaida network in Afghanistan. Other factors that have been analysed as contributory include ethnic and cultural conflicts related to niche settlement patterns in mountains; political and economic marginalization of mountain people; and hardship-driven migrations both to and from mountainous regions.

Sustainability for Mountain peoples

Mountains, even in their most pristine forms, constantly undergo natural change—uplift, volcanic eruptions, erosion, shifting vegetation, species evolution and extinction. To this, a number of human induced changes have been added—forest conversion, mineral extraction, hunting, are local changes that can change water flows and siltation downstream, eliminate both animal and plant species, and affect local risks of flooding and landslides. In addition to such relatively local impacts, global influences, such as climate warming affect mountain processes - glacial retreat, rising tree lines, etc. that have local as well as transnational impacts.

Many of these activities have contributed to income generation and growth overall. What is important is that change due to human activity does not undermine the ability of mountain regions to sustain the flow of vital mountain services indefinitely, while insuring that mountain people benefit appropriately from the good and services provided, which has often not been the case in the past.

It is important to recognize that sustainability concerns cover very different time frames. Mountain ecosystems are fragile, and can degrade rapidly in certain circumstances. Natural risks and hazards are unusually prevalent in mountain regions: in Yunquay, Peru, 17,000 people were buried within a few minutes by a massive landslide triggered by an offshore earthquake hundreds of kilometres distant; Soyuz Lake in Tajikistan was formed by an earthen dam created by mass wasting, creating a hazard that threatens downstream communities and contributes substantially to the drying of the Aral Sea; Mount St. Helen's in the U.S. Pacific Northwest erupted destroying thousands of acres of high quality forest cover; and volcanic eruptions from Mounts Etna in Italy, Pinatubo in the Philippeans, La Soufriere in Guadaloup and Martinique, among others, have forced evacuations and economic disruption, and in many cases tragic loss of life.

Large areas of European mountain forests were cut, and have not grown back due to changes in land use and soil loss. Mountain areas in Africa have been stripped of vegetation by overgrazing and are no longer capable of supporting sustainable livelihoods at current population levels and intensity of use. In many parts of the world, deforestation in mountain areas has contributed to lasting changes in land productivity. In other cases, change and the impacts of change can spread over a relatively long time frame from a human point of view.

For example, glacial retreat due to global warming is likely to occur over a period of 50-100 years until nearly all mountain glaciers are melted. During this transition period, downstream flows are likely to increase, encouraging increased dependence on those flows, which will diminish rapidly once the glaciers are gone. Land conversion (deforestation) and species depletion and extinction can easily be spread over time spans longer than a normal human life, so impacts may not be perceptible. Other mountain attributes only change over a much longer period of time, through gradual erosion or uplifting due to tectonic processes. Furthermore, like many other environmental changes, some of the changes in mountains may be irreversible in any human relevant time span, such as restoring glaciers or reforestation in higher altitudes where trees grow slowly.

Sustainability means finding ways to manage mountain resources and systems so that they can provide their critical goods and services indefinitely into the future. The opportunity cost of foreclosing the longer term availability of renewable resources must also be taken into account. Since such future values cannot be known, conventional discounted present value calculations are not always feasible. Nor can such calculations account for irreversible losses of resources critical for human use.

The objective in all of this is not to stop change in mountains, but to manage mountain resources in ways that provide sustainable livelihoods for mountain dwellers and the goods and services valued in lowland areas in ways that protect the long-term capacity of mountains to continue to provide such goods and services.

Mountain goods and services bring value on local, national, and global levels. Some of these goods have market prices when they can be converted into commodities or services that can be traded. Others provide benefits that are valued, but not priced in markets because they are not readily converted into tradable goods and services. When these goods and services were plentiful, they could be had for the taking and no prices were needed. As they became scarce, various systems were developed to manage their use and assure their continued availability.

For other mountain goods and services, appropriate mechanisms have yet to be developed to ensured their sustainable use and continued availability. Many of these goods and services are not readily converted into discrete items that can be priced in markets, but they have value for individuals, for communities, and for the functioning of economies. Fresh water is perhaps the most important product of mountains for economic well-being.

Mountains provide fresh water for over half the world's population directly. Not only is water provided, but it is usually clean and safe for human consumption in its natural state. In addition in normal circumstances, mountain environments even out the rate of flow of water between wet and dry seasons and reduce siltation downstream. Where upstream activities, such as forest conversion, disrupt normal flow and siltation patterns, this can cause significant costs to downstream users. Clean and regular water flows is being increasingly recognized as a valuable service worth paying for. There are growing numbers of cases where the downstream users are finding ways to compensate upstream users to take actions that mitigate the negative downstream impacts.

Protecting water resources can be important for local use and for national consumption. Forests are probably the second most important good provided by mountains, with varying distributions across mountain regions. About XX% of the world's timber comes from mountain regions, and the share is likely to increase as lowland forests are depleted. Timber is a mountain resource that is readily converted to a marketable commodity through logging, and it is at least notionally priced for it log value. Since most logging is done in natural forests, little is anything is paid for the tree production costs, which should determine the minimum price, together with the costs of cutting, extraction, and processing.

However, standing timber also provides valuable services - stabilizing water flow, protecting biodiversity, providing amenities, generating other products. Studies in a number of areas have placed economic values on these services and they often exceed the value of the timber extracted. Because many of these services are not traded, they lack market prices, and the timber extractors are not charged for their destruction of other valuable services.

Timbering often disrupts local cultures and production patterns. It may provide short-term income at low wages for loggers, but as an area become logged out, the timber companies move on and leave the local people deprived of their logging wages and stripped of their traditional sources of livelihood in the now depleted forests. This often leads to more destructive cutting of remaining wood, and conversion to other uses, which are often not sustainable.

Unless the forest dwellers control their own resources and gain the revenue from their exploitation, timber generally does not raise mountain incomes and often contributes to impoverishment of mountain people over the medium term. It is the downstream exploiters who benefit, usually over the medium term until wood is depleted in a given area. Restoring forests in mountain areas can be a long term proposition, if it is even possible.

Minerals are another major good provided by mountains. Like timber, minerals are distributed unevenly among mountain areas and usually extracted by enterprises from outside mountain areas who gain access to the resource for little or no payment to the mountain people. Their payments to governments are based on taxes and royalties and rarely returned to mountain areas. Ores enter markets once they are processed, converted to commodity metals, priced, and sold. The impacts of mining on local mountain areas are more localized than logging, but usually more intense. Tailings and runoff from mines can cause severe damage to nearby land and toxic pollution to water sources.

The value of the lost clean water and health damage are rarely included in the costs of the mining company. Mining can offer employment to local people in mountains, but it is often highly dangerous and risky work. The introduction of the mining economy severely disrupts traditional societies over relatively long periods and leaves only wasted lands when a mine is closed. An operator has no market incentive for remediation once the minerals are exhausted, unless regulatory systems are effectively enforced, with penalties for noncompliance.

Local communities have often revolted against the exploitation of mining, and such cases are becoming more common. In some cases, protecting employment of miners is used as an argument for subsidies to mining activities, but the miners are rarely the real beneficiaries of such subsidies.

Biodiversity and amenity benefits are among the most widespread of mountain values, and among the most difficult to assign market prices. Although there are specific species (charismatic megafauna) that attract attention on an individual level, and other exotic species that have high value for collectors, despite CITES, most biodiversity and amenity benefits stem from the integral functioning of mountain environmental systems.

These integral systems also provide important and sustainable sources of livelihood for mountain dwellers. Revenue may be earned from sustainable use of forests and other mountain products, sometimes from revenues from tourism and recreational uses, sometimes from the more healthy and satisfying environment of the mountains themselves, and sometimes from sacred and inspirational value to local people or visitors. It is increasingly recognized that integral mountain systems have benefits of global value.

Protecting biodiversity is important for preserving genomes for food crops, for developing new medicines, and aesthetic values worldwide. Preserving mountain forests and related vegetation systems intact reduce risks of landslides. Mountain forest areas can also be important for sequestration of CO2. It is very difficult to translate these benefits and values into market prices and transactions, but there have been some interesting initiatives with some success. While there are often short and medium term aspects to these actions, they typically run over the longest terms, and public interventions are more likely as markets rarely function well with such horizons.

Many of the most important mountain contributions take the form of public goods. By 'public goods,' means those goods and services that do not satisfy the criteria of private goods in that it is virtually impossible to exclude others from the benefits of a public good and use of the public good does not reduce its availability for use by others.

Most of the beneficial environmental services provided by mountains - water management, biodiversity, weather contributions, cultural, recreational, and amenity values have important public good aspects. These are often mixed with private good aspects, which makes it more difficult to determine how to assign ownership, prices, and manage them, but progress is being made. When population pressures and economic pressures are low, the degree of use of public goods does not often pose a sustainability problem. As pressures have increased, overuse and abuse of public goods becomes more of a problem.

Many mountain goods and services are part of, or the result of interactive systemic relations. They cannot be separated from the fabric of the mountain ecological systems. Water management depends on the kinds of forest and ground cover, the extent of roads, the kinds of construction on waterways, etc. Similarly, protection of biodiversity depends on changes in land use, demands for rare species, competition for food sources. As a result, activities which extract resources have systemic effects beyond the creation of commodities that enter private markets and direct impacts on the mountain environment.

Logging can have major impacts on water flows and siltation. Mining can have an impact on downstream pollution. Roads built to take out those products can have a major impact on mountains, opening up areas to further settlement, increasing erosion, disrupting ecosystems. As a result, management of mountain environment requires more elaborate consideration of secondary effects and their systematic impacts than is often the case in lowland areas.

Poverty in Mountains

The characteristics of mountains have lead to their

political power and dominance. Their resources, the security provided by their geography, the strength of their warriors, their situation astride important trading routes, and the lack of a number of diseases contributed to their domination of lowland areas. With the advent of the industrial revolution and modern economic structures, the relative advantages of mountains as centers of power have been eclipsed, and they have been marginalized.

Extensive sedentary agriculture in lowlands, manufacturing based on economies of scale, easier transportation and trade, and the broader reach of common language and culture in lowland areas enabled them to grow much more rapidly than mountain areas in modern times and become predominant. In nearly all parts of the world, mountain areas came to be regarded as backward, their people and cultures inferior, and their resources fair game to support lowland economies. Population densities in mountainous areas can be relatively high, despite the difficult terrain. This may be due to population, economic, or political pressures from nearby lowland areas.

Mountain land is usually less productive, and the mountain people rarely receive full benefits from resources extracted from mountains. Their relative isolation reduces their voice in political decisions. As a result, lowland governments often applied lowland based institutions of governance and resource management on mountain people, usually to their disadvantage. Control of most of the value of natural resources tended to be appropriated by governments or lowlanders, and mountain people became dependent on wages from the exploitation of the natural resources for their livelihoods, while the asset values and rents tended to be allocated elsewhere.

Difficult terrain in most mountain areas mean that constructing physical infrastructure is relatively expensive and yields low social returns, except where the infrastructure serves the exploitation of natural resources. Mountain areas are isolated in physical terms and until recently in terms of communications. Similarly mountain areas tend to be poorly served with social and human development infrastructure.

Mountain groups often have different ethnic and cultural backgrounds from the lowland groups who control political power, further reducing the voice of the mountains and contributing to the lack of interest and resources from the lowlands. This marginalization is common in both developed and developing countries, accentuating difficulties that result from lack of access to centers of population and markets. These factors help explain why mountain areas have a much greater incidence on poverty and lower levels of development almost everywhere. There are a few exceptions where countries have recognized the importance of mountain areas for their environmental, economic, and amenity values. This tends to come at higher levels of development.

Switzerland, a highly mountainous country that created most of its wealth in lowland activities does value its mountain areas for their environmental and amenity values and as representative of the image of the country. As a result, it has worked hard to maintain and enhance its mountain areas through a variety of programs that recompense mountain people for the services the mountains provide and to maintain the tourism value of the mountains.

In parts of the Rockies and other mountains in the US and Canada, the amenity and tourism value of mountains has led to renewed interest and investment in mountains and preserving mountain environments in those regions where the amenity value of tourism and recreation exceeds that of disruptive extractive industries such as mining and timber. Elsewhere, mining and logging tend to prevail in the absence of direct market competition from tourism.

The overall result is not the lack of intrusion, but different types of intrusion and different impacts on the mountain environment. People in mountain areas tend to be fragmented by geography, and often by language and culture. Social and political institutions are usually quite localized, and often quite strong within the local area. Although mountain people are independent, the rigours of life in those areas requires strong local social systems to assure survival. The relative isolation tends to make such local groups distrustful of outsiders, and particularly of non- mountain people.

Fragmentation and lack of communication with lowland power structures contributes to the isolation of mountain areas and their poverty. The key to improving standards of living in mountains is improving local institutions to give local people more control over mountain assets and the means of negotiating more equitable allocations of their benefits. In part this will involve improving the provision of education and health services so that mountain people will be able to mange their assets better in the modern world. This will in turn depend on building more equitable relations with the political institutions of the lowlands and assuring a more equitable distribution of public services. Such a rebalancing of power should be based on greater recognition of the importance of services mountains provide and the expanded role that mountain dwellers can play in assuring the provision of these services. A certain degree of enlightened self interest on the part of lowland institutions as well as well coordinated actions on the part of mountain people will be required progress in this area. It is particularly important to obtain recognition for community property rights in mountain areas and vest control in local institutions through the application of principles of subsidiarity.

Means of identifying and quantifying the value of economically and socially beneficial services provided by mountains will permit creation of appropriate institutional structures, and proper remuneration of mountain people for services, thus improving the quality of life in both mountain and lowland areas. As will be discussed below, market mechanisms can be used in many instances, once the proper institutional framework is in place.

Means of improving communications and trust between upland and lowland groups will be essential to make progress in these areas. It is clear that there is no one-size-fits-all approach. Different methods will have to be used in different areas and adapted to local circumstances. However, it is possible to identify a well defined range of possible situations and appropriate solutions as a starting point for local adaptation.

The populations varies depending on their natural and community resources, their knowledge of resource use, their access to lowland markets, and the lowland valuation of mountain goods and services. The latter two factors in particular are often related to the degree of political access and voice available to mountain people as a recognized political constituency.

Natural resource endowment, sometimes called "natural capital", provides rural populations with the most critical wherewithal with which to nurture themselves and their families. Complex regimes support the maintenance of such systems, involving an intricate web of legal, economic, social, cultural, and often religious incentives and sanctions to ensure that the resource base is maintained for present and future generations. Where highland producers have access to markets for high value added products, or where they receive transfer payments or premium prices from lowland markets in return for environmental services, such connections can help mountain communities to achieve a good quality of life. It is the absence of one or another of these factors that creates special problems for conservation, poverty reduction, and sustainable development in mountains.

Sustainable development approches must be locally adapted. Nowhere is this moie true than in mountains, where variations of altitude, rainfall, insolation, temperature, soil, vegetation and other factors create myriad micro-climates. Replication of "best practices" must take account of the exceptionally high variability of mountain ecosystems and cultures. Nevertheless, it is possible to construct a framework for assessing different mountain areas and circumstances in order to help design the best approaches to contribute to sustainable development and poverty alleviation.

Almost all commercially developed hydropower originates in mountain regions, and a disproportionate share of mineral and timber resources also come from mountains. For timber, this was not always the case; but easy access and higher land values for agriculture and other uses led to deforestation of many lowland forests over the several centuries, so that much commercial forest land is now in mountainous regions. Many minerals are common in mountains because the geological upthrusts and folding bring veins of metals and minerals close to the surface in mountains, making them more accessible.

Natural resource values cover a vast range of possibilities: biodiversity; maintenance of watersheds, traditional knowledge and production systems, environmental amenities such as recreation, aesthetic and inspirational value, and the powerful appeal of untrammeled nature. The service and amenity values of these resources depend on the recognized benefits conferred on downstream dwellers. The higher the appreciation of these services and the higher the income levels of downstream inhabitants, the greater economic value likely to be assigned to these services, and the greater the potential for viable economic transactions to help preserve these environmental services.

Mountain ecosystems also provide carbon sequestration from intact and regenerating vegetative cover, risk abatement through mitigation of landslides and other hazards, and generation of weather regimes on which current patterns of development are based.

Mountain cultures have co-evolved over long periods with the ecosystems in which they exist, leading to development of detailed indigenous knowledge systems, local stewardship of timber and non-timber forest products, high altitude production systems, and specialized food and craft products. Maintenance of such systems is critical to the sustainability of local mountain communities, who are the stewards of upland resources; and it is also critical to continued availability of specialized products that are highly valued by lowland populations.

Mountains are the second highest vacation destination for tourism. This tourism generates income for mountain dwellers and provides incentives to preserve the features of mountains that attract many visitors. Such mass-producedtourism, however, often increases environmental degradation more than it benefits local economies. Extensive visitation can reduce the amenity values that attract tourists in the first place. Recreational and aesthetic values, sacred and inspirational values, and the desire to experience nature in a pristine state are strong factors in the exceptionally high amenity value of mountain regions.

There are often competitive demands on mountain resources - increased resource extraction reduces the extent and value of in-place services, or preserving in-place services may reduce what can be exported. Furthermore, the relative value placed on these resources depends on the tastes and technology of the rest of the world's economy. The values of export commodities are determined by world markets, and often by the tastes and income levels of the developed countries.

In developing countries, this creates a bias for export uses that may or may not be justified depending on the use of the proceeds of the exports and calculation of the in-place values. The fundamental differences in the economic or social value of export commodities and in-place services and natural resource value are further affected by differences in access to markets. Easy access to markets makes export uses easier, but may also place a higher value on in-place values. On the other hand, lack of access to markets may help preserve in-place services without having to create an economic transaction, but it makes it more difficult to alleviate poverty. Access to markets represents the major factor in how much economic value can be realized from both export and in-situ resources and in determining the sustainability of their use. It is also an important factor in alleviating poverty in mountain areas. For this analysis, access to markets consists of three complementary elements:

1) Investment in transportation infrastructure to improve physical market access;
2) Application of appropriate technology to mountains to provide communications linkages (wireless telecom), to provide utilities (solar power, mini-hydro), to improve knowledge, (distant learning, internet access), and to enhance the value of mountain products; and
3) Development of institutions to improve the sustainable management of mountain resources, including more participatory governance, application of market mechanisms to a broader range of mountain goods and services, and appropriate regulatory interventions.

In mounains, cultural and linguistic commonality will also increase the likelihood of high access. On the other hand, greater distance, more rugged terrain, and cultural differences are likely to reduce physical access and access to markets. Many of these factors can change over time and are affected by technology, policies, and institutional innovations. Lowland tastes and technology may change, which can increase or reduce the demand for mountain resources. Higher incomes are likely to increase the demands for and relative valuation put upon in-place services. Migration in and out of mountains is likely to increase access.

Conflicts and illicit activity is likely to reduce access, such as the Shining Path in Peru, or the Maoists in Nepal, or the drug dealers in the Golden Triangle. Roads and other modes of transportation are the main physical form of access, and impact directly on the capacity of mountain services by their existence as well as the opportunities for linking to markets. In recent years and in some locations, air travel and telecommunications have opened up a wider range of access to mountain areas.

In some cases these are less disruptive than roads, and may offer opportunities for income gains from in-place activities. Given the threatened and fragile nature of many mountain areas, the type and rate of expansion of access is important. In some areas, it should be limited, in others expanded. Enhancing some in-place environmental services may depend on increased access to market transactions to monitor and remunerate stewardship activities, but controlled physical access. The discussion of the various cases below will highlight how different approaches are needed in different areas.

The most intractable problems of sustainability are likely to arise in the "High-High" (high economic and high environmental value) and the "Low-Low" (low economic and low environmental value) cases. The Low-Low case represents the typical form of mountain marginalization. There is little for market interests to go to that area, and the people are most likely to be poor due to lack of resources and contact with the rest of the world.

The High-High case represents the typical case of conflict and competition for mountain resources. Without careful management of both types of resources and reasonable compromise, sustainability is truly at risk. The cases of High export resources and Low in-situ resources (High -Low) and Low export resources and High in-situ resources (Low-High) are likely to be more manageable due to less conflict over sustainability objectives. Nevertheless, they require careful management to assure both sustainability and assuring adequate incomes to the mountain dwellers in those areas.

Such areas are of high concern for social and economic development as they represent the extreme of marginalization. Poverty is pervasive in these areas, which have limited capacity to provide incomes from local resources or from export of mountain products. Populations are likely to be ethnically or culturally isolated from nearby lowlands, further adding to isolation and have only relatively weak institutions beyond local clans.

Environmental risks are likely to be due to neglect of potentially important issues. People in these areas are particularly vulnerable to the fact that caloric requirements of living at higher altitudes are greater than in lowlands, given metabolic difficulties of thinner air and colder temperatures, while agricultural production is decreased due to altitudinal conditions. These mountain areas are frequently characterized by environmental degradation. This may have been due to excessive population pressure and migration from nearby lowland areas that depleted or exhausted fragile resources, perhaps resulting from past mismanagement.

Population centers may also be along isolated trade routes, where the population would support and depend upon supporting the passing commerce. The primary challenge in these areas is survival and basic poverty alleviation. The cost of providing public services and other poverty alleviation is high in relation to any immediate economic return. There is little immediate benefit from improving living standards in these areas beyond helping the local people.

Lack of resources increases competition for the sparse resources that are available and may make developing a coherent program difficult, particularly if there are ethnic or cultural differences between the people in the isolated mountain areas and the centers of wealth and power. Policies would have to center around welfare programs to supplement incomes and improve social and educational services; migration; and improved technical packages.

Some areas in this quadrant are characterized by high altitude urban agglomerations, such as Mexico City and La Paz, which are highly or entirely dependent on their lowland hinterlands. They may have been located there for historical reasons, and their growth due to their position as national capitals. They impose a heavy burden on their local environment, such as water depletion and air pollution and draw on resources of the rest of the country. In these areas, there is little left to preserve of the original environment, and decisions will be made primarily on the needs of the local populations.

However, improving general environmental conditions are important for improving the welfare of the inhabitants. The primary challenges are managing the environmental conditions that result front the intensive use of these areas. This involves control of pollution, ensuring adequate and safe water supply, and managing population expansion. Providing clean air and water will depend on applying known technology wherever possible, imposing pollution control regulations, and undoubtedly passing on the costs to the bulk of the population through higher prices, taxes, or user fees. Care should be taken not to adversely impact the poor in these programs. In addition, national- level policies would be needed to limit population growth and in-migration while encouraging out- migration.

In areas where there is less population density and where access to markets can be enhanced, it may be possible to develop environmentally benign income generating activities in manufacturing and services. These programs would have to be carefully designed to protect the valuable resources and reduce degradation.

At the same time, they would need to develop better opportunities for the local poor. There may be some possibilities for improving management of environmental services, such as water. This might make it attractive to develop some market based programs to pay local people for stewardship to prevent downstream damages, if they have access and control over the causes of the problems.

However, payments based on stewardship services are not likely to be significant. If the main source of environmental degradation is exploitation of the meager remaining environmental resources, it will require stronger interventions from the government or external agents to establish guidelines for better behaviour. Where market transactions can be arranged that directly benefit the poor, they should be used. Beyond that, increased social services should be funded by the government.

The potential for developing income generating activities and environmental services would depend on developing strong institutions and appropriate relations with downstream markets and beneficiaries. The Zuni native Americans live in a Low-Low region, but can make a reasonable living by selling their beautify jewellery and other crafts in nearby affluent markets. Tibetans make equally impressive jewellery and crafts, but do not have an affluent market nearby. Establishing such relations to generate incomes may encounter political problems, depending on the circumstances. These factors would also affect any potential for migration.

The extreme harshness of the Tibet Autonomous Region of China, for example, makes its biodiversity value relatively low, and its isolation reduces the value of other environmental resources. However, the critical role of this Himalayan as the source of many of the world's great rivers, its mineral potential, and its cache as the cultural home of Tibetan Buddhism are increasing recognized by key agents in these markets and shifting it from low to high natural resource value for both exports and in-situ benefits. The valuation of these resources is based on market perceptions of scarcity, impacts on other (downstream) activities, and commercial benefit. These change over time. Demand for these resources may contribute to increased access for TAR, with the attendant benefits—and risks.

In addition, as downstream areas from certain isolated areas become more affluent, they may place a higher value on the resources from nearby mountains. Or such areas may become valued due to their location in relation to other activities, such as the potential value of the Pamirs or Caucuses as a means of transport for oil. In areas with both low value environmental resources and low economic potential, offering credit to help boost local economies is one promising mechanism to alleviate poverty .

Low environmental services value requires careful management of the exploitation. These areas feature relatively high value commodities, with relatively low values for in-place environmental services. Many of these areas are found in the Low Mountainous areas, and to a lesser extent in the middle levels of the High Mountainous areas . In these cases, the potential for conflict among uses is less severe, but not without challenges to protect the environment from pollution and degradation and to assure that local people realize a fair income and long terms opportunities from the exploitation of the mountain resources. Where these resources are accessible to markets with appropriate infrastructure, there is good potential for sustainable use of the resource and reduction of poverty.

In most developing countries, the bulk of the mountain products is exported into international markets. In many cases, the right to exploit the resource is determined by national governments, who normally claim ownership of such resources. The benefits of exploitation rights are typically given to large, well funded firms—both foreign and domestic-who exert influence over government regulations to their own benefit. Revenues are usually appropriated by intermediaries and lowland elites who gain ownership of the resources. Local mountain people are rarely compensated for the resource, or for the loss of other resources that results from the extraction activity. They may be employed in menial jobs in the mines or cutting timber, but rarely is their standard of living improved.

Extractive firms have an incentive to minimize costs of the mine operation and the transport system, and may try to get the government to contribute the infrastructure, which would divert public resources from other uses that may improving living standards in mountains. In the case of the Antamina Mining Company, a collaborative local effort involving the help an international NGO resulted in an alternative transport system that avoided an important National Park as well saving investors money in the long-term.

Because of the public good aspects of many of these goals, a combination of market and regulatory regimes would be necessary to protect the environment and the access of local people to the resources of their mountain areas needed for their livelihood. However, once an overall framework is agreed upon, with adequate participation from the local people, it is likely that many market-based mechanism can be used to implement the policies.

For example, the investment of funds from the downstream purchase of hydro-power into upstream stewardship programs in Costa Rica in one way to provide stewardship incentives. Where reasonable institutions and social capital exist or can be developed, a range of effective policies can be supported. Where mountain areas possess high value products, but have low access to markets, it becomes important to manage the construction of the necessary infrastructure to enable the evacuation of the product with minimal disruption to the environment.

Mining is the most common example, as the resource is often discovered in remote mountain regions, and requires appropriate infrastructure to transport the product at commercially viable costs. Logging for timber and pulp and paper may fall into this category as well. In these areas, it is important to prevent the extraction of resources from damaging other environmental services and sources of income of local populations. This requires government regulation and appropriate enforcement for pollution avoidance and mitigation of degrading effects.

In some cases, such as forest areas, local populations can be more involved in the management of the resources, using traditional social and environmental approaches. Depending on the situation, it maybe desirable to limit the physical access to the site to the extraction of the resource and discourage expanding access. This would be true where the area has a fragile environment and where allowing expanded settlement would have adverse social or environmental impacts.

Furthermore, people living in these areas are often poor and in need of means to improve their standards of living, which requires greater access. The challenge is to conserve these areas in an affordable manner and to keep physical access limited. Environmental impact assessments have been completed and basic mitigation measures undertaken to reduce the direct negative impacts of the export of resources.

In the Low-Low instances, the primary concern will be on improving the welfare of the mountain people living in those areas. This will primarily involve government sponsored programs for subsidies and payments, but other instruments may be used as well. The emphasis should be on poverty alleviation. In the High export resource, Low in-situ resource instances, the primary concerns will be to mitigate potential damage to environmental sustainability while assures that the local people receive adequate compensation for their resources and their exploitation.

The High-High instances are intrinsically complex and involve difficult trade-offs. Some combination of the policies for the High-Low and Low-High cases would be required to protect the long term viability of both the export and in-situ resources while assuring livelihoods for the local people. Cooperation of public and private institutions will be necessary, using market instruments to the extent possible, but recognizing the role of the public sector in assuring the public goods are adequately valued and managed.

Land rights are the basic, if not always the definitive, factor in the use of mountain land. Traditionally in most mountain areas, use was open access until various pressures on the land created the need for more control. Usually some form of community ownership evolved, where the primary group using a mountain area allocated use of various aspects of mountain resources to people in the community. This approach preserved the integrity of mountain resources, helped mitigate risks and variability of output, and managed the resources to meet community needs, while excluding outsiders.

Combinations of more intense pressure on the resources and the introduction of national government authority led in many areas to the states pre-empting land rights. Different states have then allocated use rights in different manners. Some land has been put under conservation rules, some has been converted to private property under a variety of systems, and has been licensed for specific uses - all with varying degrees of enforcement. In most cases, these changes have not led to more sustainable use of the environment resources, nor to poverty alleviation in the mountains.

Clear and proper definitions of land rights are most important in the High export value cases. That is where the terms of access to the exportable resources has to be defined and the controls on their over-exploitation imposed. And that is the source of most contention. With low-export values, there is more flexibility in the treatment of rights, and in the low- low case, this is much less important a factor than providing adequate standards of living and human services.

While such policies have been criticized for being overly cumbersome and inefficient, there are cases where they are the only option. Such interventions may be the only way to prevent degradation and protect mountain people's livelihoods. This may include prohibitions on release of toxic chemicals from mining, controls on road building activities in delicate areas, etc. Obviously, it is important to design such regulations carefully and be sure they are enforced. These policies will usually have the greatest impact on high in-situ value cases to control and manage the public goods aspects until adequate proxy markets can be established.

Taxes and fees are traditional government mechanisms to raise money and influence behaviour. These can be used to pay for stewardship services where the public good nature of the benefits do not permit more market oriented approaches. Taxes can be imposed on pollution emissions, to mitigate degradation, or imposed on downstream users to pay for control and mitigation activities. Fees are often used where there is a specific service used, such as entry into national parks. The may also be imposed for the extraction of a natural resource, such as timber and mining. What is important here, as with creating markets, is that the taxes and fees collected by used to support stewardship and better management of the resource - staffing and managing national parks, controlling extraction and appropriate intermediation, such as reforestation.

Public payments for stewardship of public goods should reflect general agreement on the value of such stewardship which justifies the payments, either national or internationally. These payments should be market value based to the extent possible, but they may include some additional transfers to combat poverty or increase social services. Taxes and fees are best imposed on items of value where there is a willingness to pay. They also have to be designed in ways that minimize distortions. In high in-situ resource situations, they can sometimes be used to substitute for market pricing of a public good, e.g. taxes on the emission of a pollutant or for entry into sensitive area.

Transport infrastructure is very important for improving access of mountains people to markets and other benefits of lowland areas. At the same time, the location and management of infrastructure can have critical impacts on the development of mountain resources. Road can open up resources for export, and they are a major source of erosion and degradation. Roads can also open up areas for settlement and exploitation. The pattern of infrastructure development is a major tool governments have for managing mountain area development and should be planned carefully. Private interests that want to exploit mountain resources may build their own roads, or try to get governments to do so for them.

In some cases this can be quite damaging to the environment. And as in the case of Antimina in Peru, it turned out that alternative methods was more economic in the long run, but only adapted after pressure from national and international sources. These activities are important in all areas of limited physical and communications access. Since the design and construction of the infrastructure can have either positive or negative impacts on the environment and the well- being on mountain people, it is important that the infrastructure be sensitive to these issues.

In particular, governments should consult locally on infrastructure needs and refrain from building roads simply to facilitate the export of mountain resources as a subsidy to the exporters. Provision of social services are another essential role of governments that play a critical role in the alleviation of mountain poverty.

Better education and health increases the capacity of mountain people to improve their standards of living and undertake new income generating activities. In many cases, the combination of formal education with local traditional knowledge can greatly enhance the ability of mountain people to participate in stewardship transactions. Some of these are sophisticated and require higher levels of understanding than may be found in traditional areas.

Education can also enable mountain people to practice better environmental protection in their normal activities and identify and develop high-value niche markets (e.g. Switzerland). And it can equip them to gain livelihoods from migration where mountain areas are too poor or densely populated to support their populations at adequate levels. This is most important in the low- low areas, which need better social services and are usually the most deprived. Providing these services is justified on a equity basis and as a means of improving the prospects of people in these areas to increase their standards of living or to migrate. These services are of course important in all areas and essential for building the capacity among local populations to better manage their affairs and engage in more productive market transactions as the opportunities arise.

Transfer and welfare payments are also part of a government's toolbox. They are often abused, but are an essential last resort for areas where income opportunities are meager and migration not feasible. These payments could be associated with basic stewardship activities as well. Programs to improve sustainable economic activities may be used to create more viable jobs and income earning opportunities. These may also contain a subsidy element to offset some of the inherent disadvantages of working in mountain areas. Such transfers could also reduce the incentives to migration or resource exploitation, which would have identifiable public benefits, even if not quantifiable.

Unfortunately, most developing countries do not have resources for substantial transfers, and usually the mountain people lack the voice to obtain even a fair share. This deprivation may lead to conflict and participation in illicit activities. This are primarily for the low- low areas who are most in need. They may be of value for certain deprived groups in other areas as well.

Local participation is a very important means of involving local mountain people in issues of importance to preserving mountain environments and improving their well-being. It enables them to express their voice and to strengthen community management of their resources. In some areas, local communities have come together on their own to institute better community management with the acquiescence of the national government. In some cases such activities are actively encouraged, and unfortunately in other cases, government discourage such local initiatives. Where local participation is strong, there are generally positive impacts on both environmental sustainability and standards of living, as people take responsibility collectively for their own long term interests. The capacity to engage in varieties of local participation is very important for organizing market based stewardship activities and for assuring equitable distribution of returns.

These institutional developments are most important in areas of high in-situ values and provide a means of organizing local groups to recognize the value of the environmental services and to better realize their value through market or other interventions. Participation is important in all areas as a way of getting people to work together to express and eventually achieve their goals. Local knowledge and local commitment have proven very important in motivating local success.

Identifying and quantifying environmental goods that do not have the usual private good nature of normally traded commodities is a crucial first step. A tree is readily identified as a commodity - timber. It is not nearly so readily identified as an economic good when it is part of a system which regulates water flow, manages siltation, or provides recreational services. Because of the public good aspects of these services and the fact that they cannot be segmented into individual chunks that can be traded in a conventional marketplace, it is much more difficult to conceive of them as goods which can be treated as marketable. Recent advances in the understanding of the role and functioning of ecosystems in providing valuable services and identifying clear benefits from environmental systems have found ways of quantifying these services in ways which allow use of market based systems.

Since these environmental goods are not separable, various virtual systems have been developed to measure the benefits, roughly assign amounts to various beneficiaries and attribute 'production' to stewards. Establishing market systems for many of the environmental stewardship systems prevalent in mountains is more difficult. Most to the services are the result of maintaining the integrity of mountain ecosystems, preventing degradation, and restoring degraded areas.

4

Impact of Biodiversity on Livelihoods

Today, Bioresources play a critical role in the livelihood systems of many of the world's poor, even in highly modified or degraded landscapes. By contrast with the ecocentric paradigm which suggests that all living species have an intrinsic value irrespective of any value that people derive from or attribute to them, the anthropocentric paradigm views biological resources as the collection of 'goods and services' that support human life. In this case conservation is necessary where depletion or degradation threaten future stocks of natural capital or endanger ecological functioning and life-support systems.

The management of bioresources by the poor is of central relevance to questions of poverty alleviation and policy development. It is crucial that the social groups most dependent upon bioresources should benefit from conservation efforts or be compensated adequately where effects are negative. Bioresources have diversity and abundancy attributes that provide a range of goods and services of benefit to people, and it is these goods and services that drive anthropocentric arguments for conservation. In economic analyses these goods and services are generally divided into use and non-use values.

Most scientific or formal definitions of biological diversity (biodiversity) focus on diversity itself, i.e. the variety of different living organisms (including genetic, species and ecosystem elements) in the world or a given land area. In practice, however, the concept is commonly used in a less precise way to denote the totality of nature - i.e. not just its diversity but also its abundance and the ecological functions that it performs.

In the rural development the diversity of nature can underplay the importance of biological abundance: one bee, however rare it may be, is hardly important by itself and its value to pollination lies in the fact that there are many millions of them. In practice the focus has led to an emphasis on saving endangered species, particularly prominent mammals such as whales and tigers, rather than less tangible but vital ecological functions (e.g. the contribution made to nutrient cycling, watershed protection and biosphere resilience) or more mundane bioresources of value to local people.

Use values can in turn be subdivided into direct, i.e. the physical goods used by people (such as food, fuel, timber, and herbal medicines) or aesthetic or recreational benefits obtained, and indirect, i.e. the ecological functions that maintain the stability and productivity of the environment at local and global scales. These include pollination functions performed by insects and bees, and contributions to nutrient cycling and fertility maintenance by soil organisms.

At a wider level, trees and scrub vegetation contribute to environmental services such as soil conservation and hydrological protection and, more widely still, to regional (or global) climatic regulation. The major non-use values are option and existence values. The former relate to potential future uses such as the availability of plant material for breeding or the possible discovery of new medicines, whilst the latter refer to the values placed on the continued existence and survival of rare and threatened species and ecosystems.

Many of these values are extremely difficult to measure and use, and the values ascribed are not socially neutral. In particular, local institutions such as property rights often give rise to differential access to and use of bioresources by different stakeholders, greatly affecting assessment of value. Weeds, for example, may be a serious impediment to settled cultivators while contributing much to the diet of landless people.

The benefits derived from bioresources and the costs of their conservation can therefore be very unevenly distributed. Similarly, the values of bioresources are dependent on the level or scale of analysis. Most local values are utilitarian direct use values that provide immediate and practical benefits to local people. At a regional level, direct use values become less important compared with the indirect values of ecosystem functions and services.

At the international and global scale, non-use existence and option values become critical, though questions of biosphere maintenance are also of vital importance. As the benefits and costs of bioresources are distributed unevenly in society, it follows that the micro-economic impacts of interventions are also uneven.

The nature of impacts depends on locally-specific factors such as the quality and availability of resources, the characteristics of local institutions, and the nature of environmental-livelihood patterns and interactions. This analysis suggests the need to prioritise protection of those bioresources that provide benefits to local populations, especially the very poor. It also suggests the need to take full account of the opportunity cost of environmental change, and that the calculations are done not just at global, national or project levels but also from the viewpoints of those stakeholder groups directly concerned.

In Africa many controlled hunting areas and game reserves were reclassified as national parks after 1945. The focus was on the establishment of protected areas in places of prime importance to biodiversity or where high-profile animals or habitats were threatened. As in the early years of nature conservation in the USA, most conservation followed a fortress approach that separated habitats and people; those living within the park boundaries were increasingly excluded from use of the natural resources on which they had previously depended. This top-down and often globally-focused approach to conservation generally failed to protect wildlife as fully as intended and brought hardship to local communities.

Since the mid-1980s, it has increasingly been replaced by a community-based approach aiming to work closely with local people, draw on their knowledge, and contribute to their livelihoods. For the most part conservation is, however, still concentrated in and around protected areas, nowadays often surrounded by buffer zones in which certain identified economic activities are permitted.

The majority of rural people live in modified zones where agricultural and livestock systems predominate. In these areas biodiversity generally has limited global significance but is nevertheless of considerable use-value to the local poor. Recognition of its micro-economic values parallels the growth in conservation activities in agricultural circles in Europe where environmental conservation or enhancement has become a legitimate commercial operation.

There are number of different strategies for integrating longer-term environmental and developmental goals have recently been developed within the nexus of sustainable development in and for developing countries. The "new architecture" of aid aims to integrate poverty, environmental and economic growth policies into development frameworks and strategies owned and operated at country-level. It builds on earlier developments in participatory methodologies and involvement of local stakeholders in the design and (joint) management of programmes and projects at a local level. It also builds on progress in the mitigation and management of conflict, though it is noted that these issues are best addressed (and can be minimised or avoided) through identification and planning at the design stage of intervention

The current challenge is to find ways to operationalise this new architecture in ways that represent the interests of multiple stakeholders at global and local levels. In most cases where development or conservation initiatives are contemplated, decisions regarding matters such as the clearance or protection of forested land have to be made, implying trade-offs between short and long-term goals and conflicts of interest between stakeholder groups.

The challenge is considerable, implying the need for appreciation of numerous people-ecosystem interactions and locally-variable political, social and institutional barriers. It also implies an understanding of the economic values of unmarketed goods and services and the difficulties of assigning and utilising these values in practice. To examine these issues more deeply requires detailed local knowledge but the costs of acquiring this are often unacceptably high in developing countries. They include:

— the difficulty of identifying options for bioresources conservation that can also sustain or enhance local livelihoods

— the difficulty of taking account of locally-specific cost-benefit distributions in policy development and large-scale (national) projects and programmes

These difficulties put a premium on the development of simple, low-cost approaches for use at the design stage of interventions, to improve understanding of environment-livelihood interdependencies and the range of different interest between macro and micro stakeholders. Such approaches should improve predictability, reconcile different interests, and increase the likelihood of intervention success.

Environmental Issues

Rapid deforestation and biodiversity losses are depriving people of valuable forest resources, such as fuelwood, food and medicine. This particularly affects the poor. Moreover, environmental hazards due to unsafe water, inadequate sanitation and waste disposal, and air pollution are a major problem for the poor. The poor are particularly vulnerable to shocks arising from environmental change, conflict and natural catastrophes. They tend to live on hillsides or flood plains and in insubstantial housing and are therefore most directly affected by natural disasters. When disaster strikes they have the fewest assets to consume or sell, or savings to draw on.

The political élite in many poor countries have been the principal beneficiaries of their nation's natural capital but have failed to convert this into assets or entitlements for the majority of the population. Natural resource taxation in a context of reduced corruption can provide a major source of revenues to support pro-poor investments in future.

In Uganda, collaboration between DFID economics and environment advisers resulted in the integration of environmental issues in the budget support submission, recommending donor support for:

— the introduction of 'green' national accounts and green taxation to reflect and redress the problems of over-exploitation of natural and environmental capital;

— integration of environmental sustainability into both policy and investment mechanisms of the Plan for the Modernisation of Agriculture (PMA);

— guidance on the related issue of integrating environmental issues into the District Development Planning process;

— promotion of inter-ministerial dialogue on the need for promotion of 'environmentally-friendly' energy as identified in the Poverty Eradication Action Plan (PEAP);

— identification and use of appropriate poverty-environment indicators.

Several opportunities are open to donors to mainstream environmental issues in programme and sector aid. These include: the potential role of strategic environmental assessment and project support; environmental conditionality, 'green' taxation and accounting; and SWAps, specifically focusing on environmental issues.

Strategic environmental assessment (SEA) is a systematic process for assessing the environmental consequences of policies, plans, programmes or proposals to ensure that they are addressed early on in the decision-making process. There is also no unique methodology for undertaking detailed environmental assessments of the SEA.

In Canada and the Netherlands, staff involved in preparing aid submissions have access to an on-line guide for the environmental component. Yet this is only a partial solution: the ultimate aim is to ensure that the PRSP has taken environmental issues into account—in effect that the SEA has been 'owned' by the recipient government. This can be a major undertaking and it requires specialist skills as well as donor support in many cases.

Forestry policy reform in Indonesia has been unsustainable and has produced few benefits for rural households. However, in 1998 following the Asian financial crisis, negotiations with the IMF and World Bank produced specific commitments to both forestry and environment sector reform. Initially, entrenched opponents of reform resisted meeting these commitments.

Political changes starting with the resignation of President Suharto in May 1998 strengthened the constituency for forestry (and more general) reform. Moreover, the forestry reform coalition broadened to include civil society representatives calling for social justice and accountability and by 1999 had implemented important forest sector reforms. A new concession management regulation demanded higher royalty payments and finally instituted performance bonds with independent monitoring for concession holders.

Civil society can also have an important role in monitoring sectoral spending agreed by recipient governments. In South Africa, a local NGO has effectively monitored and publicised spending on women and children across government ministries against government pledges. This model is likely to be adopted in other countries and could be extended to spending on the environment. The term green taxation is used in a broad sense to cover policy actions that have both revenue raising and environmental objectives.

The most useful green fiscal instruments for developing countries are reducing subsidies that have harmful environmental effects, taxing pollution and levying user fees on natural resources. Subsidies that result in environmental damage also tend to benefit the relatively well off rather than the poor. Subsidised electricity, for example, misses out the rural poor who have no access to grid supply. The international evidence is that green taxation has real potential to raise additional sources of revenue for developing countries.

Collaboration between donor agency economic and environmental advisers should identify cases in which pro-environmental policies will save recipient governments money. In general, though, it is necessary to design appropriate fiscal instruments and to strengthen regulatory institutions to narrow the distortion between private and social costs of economic activities.

In 1979 China began to experiment with a levy on industrial pollution that exceeded emissions standards in cities. In the early 1980s a pollution levy system was officially incorporated into law and was gradually expanded to cover the entire country. Government revenues from the pollution levy have since increased rapidly, e.g. from 1.2 billion yuan in 1986 to 2.7 billion yuan in 1993. The pollution levy now provides about 15% of all capital expenditures for pollution control and is the principal source of funding for regulatory enforcement activities by local environmental protection bureaux.

Water supply and sanitation and indoor air pollution are the two largest causes of lost disability-adjusted life years in developing countries. Viewed from a sustainable livelihoods perspective environmental concerns also have a legitimate role in transport sector SWAps and agriculture SWAps. While it is theoretically attractive to integrate environmental considerations into conventionally-defined SWAps, there are also counter-arguments.

There may, for instance, be some reluctance to make health SWAps more complex by introducing roles for ministries of water and sanitation and energy. In such cases it is quite conceivable that donor-coordinated support for environmental education would take the form of a project designed to support a health ministry SWAP. The longer-term goal would be to integrate this project into programme or sector aid.

The livelihoods of the poor are currently worsened by: the failure to enforce the few existing pollution controls; ineffective management of environmental issues in rural areas; minimal provision of environmental services (e.g. sanitation, solid waste management) in urban areas; and little monitoring of change and people's consequent vulnerability. A lack of empowerment constrains the ability of people to improve their situation.

Aquaculture and Livelihoods

Aquaculture is a relatively new and underdeveloped farming practice compared to agriculture and animal husbandry, even in many parts of Asia. Its positive social and environmental attributes make it an attractive entry point to improve the livelihoods of the poor in rural development programmes. In parts of Asia, poor farmers farm fish where it is traditional practice; they consume or sell fish to generate income; have confidence in the technology; and have access to appropriate genetic material.

Fish is used here generically to include all farmed aquatic organisms. Land-base systems are commonly integrated with agriculture by stocking fish in rice fields and ponds. Water-based systems involve stocking fish directly in enclosures or attaching them to substrates in water bodies such as rivers, lakes, reservoirs or bays.

Water-based systems may provide an entry point for landless people and poor fishers to farm fish. 'Rural aquaculture' implies low-cost production with extensive and semi-intensive technologies most appropriate for the limited resource base of small-scale households. It is based on natural food for fish such as plankton, which is occasionally fertilised and/or supplemented by other feed.

Fertiliser and feed may be derived from on-farm by-products, at least in the initial stage of intensification, although even here formulated and pelleted feed from agro-industry are increasingly used. By contrast, intensive systems invariably depend on relatively high-cost, nutritionally complete diets. The annual growth of aquaculture has averaged some 11% for nearly two decades.

Although salmon and shrimp receive most publicity, they comprise less than 10% of global aquaculture production by weight compared with 50% for carps and tilapias which contribute most to the domestic food supply in developing countries.

Aquaculture production is also skewed geographically with Asia producing over 90% of global products, dwarfing Africa and Latin America at less than 0.5% and 2%, respectively. In China, with 67% of global aquaculture production, inland aquaculture production has increased at least fivefold in the past decade; it has only doubled in the rest of the world, implying large potential in other developing countries if constraints to its expansion were removed. Aquaculture contributes to the livelihoods of the poor through improved food supply, employment and income.

Many small-scale farmers have small land holdings in areas of complex, diverse and risk prone agriculture in mainly rainfed and undulating land on the fringes of lowlands or in uplands. Construction of a pond on these frequently environmentally degraded farms may also provide a focal point for agricultural diversification and increased sustainability, by providing a source of water.

The poor in well-endowed lowlands are often landless or near-landless; here, fish farming in common water bodies may help to reduce poverty, provided that the poor can access them. Inland and coastal fishers are usually landless and are amongst the most impoverished; their opportunities lie mainly with water-based culture systems.

Aquaculture has contributed in the past towards poverty reduction in poor societies in the few areas of the world in which it is traditional practice, for example China, Indonesia and Vietnam, and it continues to do so today. Consumers must perceive the value of fish and this must be reflected in market demand. This is most likely to be the case in areas where wild or farmed fish have traditionally been consumed. Early attempts to promote aquaculture often focused on household food security which may have contributed to the high rate of failed development initiatives, particularly in Africa.

For the poor to become directly involved in aquaculture as a farming practice, they need to own or rent agricultural land for culture in rice fields or ponds, or if landless have access to a water body in which to stock fish. Another major requirement is knowledge of appropriate technology, e.g. many small-scale farmers failed to culture fish in northeast Thailand because they stocked too small fingerlings at too high a density in ponds that were neither fertilised nor fed; fish suffered high mortality; were consumed by residual populations of carnivorous wild fish, or simply did not grow. Aquaculture comprises sequential stages of seed production, nursing and grow-out. In its early development in China and Vietnam, seed were either trapped from rivers or spawned by small-scale farmers at the household level.

Bangladesh has dynamic, private sector involvement in production, nursing and distribution of seed, involving numerous stages to transport seed large distances by rail, vehicles and on foot. Although it is hierarchical with unequal power distribution between various 'actors', there is considerable employment of the poor as it is labour intensive. However, many areas are characterised by inadequate seed supply, and donor funded hatcheries have often proved ineffective following withdrawal of financial support. Strategies to decentralise seed production and distribution have been developed recently in some parts of Africa and Asia. Small carps and tilapias can be bred simply without the need for expensive conventional hatcheries.

Local seed production can enhance poverty reduction by reducing cost, improving quality of seed as long distance transportation of seed is not required, and providing employment and income generation at local level. Seed production and distribution have the best prospects of sustainability where private sector involvement is strong, but these can also disseminate extension information, for instance via itinerant seed merchants.

The potential of rice/fish culture is more likely to be realised where there is a market for the relatively small-sized fish harvested from fields. Most rice/fish culture in China and Indonesia is to produce fingerlings to stock more intensive culture systems. Most rural aquaculture occurs in household ponds. These are widespread on floodplains, where earth is often extracted to raise the ground level of farm buildings. Various on-farm fertilisers and supplementary feeds are used as nutritional inputs.

Night-soil is used in a few countries such as China, Indonesia and Vietnam but is socially acceptable in only few societies; and public health concerns argue against its wider recommendation. Although pigs and poultry may be housed over or adjacent to ponds, livestock on most small-scale farms scavenge for their feed which constrains the use of manure. Scaled-down integration of intensive feedlot livestock with fish has often been introduced but such systems invariably collapse following withdrawal of external support because of input supply and marketing problems.

Wild or farmed, terrestrial and aquatic vegetation may be used as manure or fodder for herbivorous fish, but use is constrained by high fibre and moisture contents, high labour requirements, and limited area available on small farms to cultivate it. Most crop by-products have high opportunity cost. Water-based systems such as cage and pen culture, enhanced fisheries in large communal water bodies, and farming molluscs and seaweeds in coastal waters may be the only options for aquaculture for the landless and for underemployed fishers.

Major constraints include insecure access to water bodies, and the high cost of production of most cage culture. The NGO CARE attempted to involve groups of women in cage farming in Bangladesh but after its withdrawal activities ceased within six months as the women lost access to the water bodies. CARE is currently reassessing whether local NGOs can maintain access to water bodies after the withdrawal of project support. In parts of Asia, there are small-scale farmers and fishers culturing fish in cages. For example, in Vietnam grass carp are cultured in wood and bamboo cages in rivers and reservoirs, using grass, maize stalks, and cassava leaves and roots as feed.

Chinese carps are cultured in Nepal with minimal supplementary feed in nylon net cages with bamboo floats and frames in eutrophic lakes with abundant plankton. However, in Indonesia where cage culture developed rapidly to resettle farming households displaced by dam construction, the rural poor gained mainly from trickle-down benefits from labouring in the industry rather than controlling it, as it was taken over by the urban élites. As governments stock water bodies with fingerlings and restrict access, poor people are unlikely to bid successfully for fishing leases.

However, in some areas governments have made a commitment to long-term security of tenure for poor fishers: in the oxbow lakes of western Bangladesh, for instance the conditions are in place for self-managed groups of poor fishers to benefit from increased yields through stocking carps. Aquaculture may be one of the few options for technical improvement for poor households in coastal communities, which are among the most impoverished.

Poor fishers culture molluscs and seaweeds in Southeast Asia, particularly in Indonesia and the Philippines. More controversial is shrimp culture which is often controlled by the better-off or by private companies. However, small-scale farmers dominate shrimp farming in Thailand, a majority of whom were previously either rice farmers or fishers, indicating that shrimp culture has contributed to the welfare of the poor directly as well as indirectly through diversification of employment opportunities in coastal areas.

Vietnam provides another example where shrimp are now produced on a sustainable basis by poor farming households on small holdings in some areas. Aquaculture has considerable potential for expansion of area. Potential also exists for use of irrigation systems and of land unsuitable for agriculture such as saline areas and swamps.

Poverty elimination is primarily the responsibility of national governments which must be committed to, and provide an environment for, enabling poor people, who are socially excluded, vulnerable, and have limited access to productive assets and resources. The poor need to be specifically targeted with a package of support that includes training and microcredit as well as 'baskets' of choices of appropriate technology. Major national level constraints are the weak implementation of policy; insufficient knowledge relating to rural aquaculture at all levels and the limited capacity of national institutions to function as service providers.

Even when governments have favourable pro-poor policy its implementation is often weak. For instance, legislation on the use of irrigation water may unjustifiably favour agriculture over aquaculture; policy may also prohibit conversion of rice fields to fish ponds due to misplaced emphasis on grain for food security. Most aquaculture professionals and service providers currently have a technocratic and fisheries biology worldview which focuses on maximising biological yield rather than meeting local objectives, high-value species rather than low-cost food fish, and commodities rather than communities.

The major gaps in knowledge include the social and environment/resource aspects of rural aquaculture; the actual and potential contribution of aquaculture to sustainable livelihoods of the poor; and the trade-offs between capture of wild fish and culture of farmed fish, and between agriculture and aquaculture, including the use of fish in IPM. Strategic research is required on indigenous species, genetics, seed quality and disease. Research is required on potential new species for rural aquaculture, especially for areas where there are concerns that the introduction of exotic species might adversely affect local biodiversity. Breeds of farmed fish are mostly wild types in terms of genetic make-up. The poor, especially women and children, might be disadvantaged if lower value local strains currently consumed in households and at village level markets are replaced by higher value, more marketable fish.

During the end of 20th century, world fisheries improved a technology. From the 1970s, however, production from capture fisheries, as recorded by the Food and Agriculture Organisation (FAO), has remained more or less on a plateau of 80-90 million mt.

In marine fisheries, there has been something of a shift among the major players. In recent years, developing or emerging countries have been taking an increasing proportion of the catch, with China and India featuring increasingly, so that now these countries take more of the world catch than the developed nations. By contrast, over the same period, there has been a sustained increase in aquaculture.

Some study shows that, global aquaculture production to have increased to 28 million mt annually. By far the greatest contribution to this is from China, South Asia and South East Asia, parts of which have an aquaculture tradition going back for more than a millennium but where modern concepts and developments have been met with a ready acceptance. By contrast, production from Africa and South America remains minimal despite, in the former case, the provision of much technical and financial assistance. Marine capture fisheries can be broadly divided into two categories:

— Coastal or inshore fisheries - most often artisanal in nature.

— Offshore fisheries - which are largely commercial; requiring mechanised ocean-going fleet for their · exploitation.

The offshore marine stocks can be fished by vessels of the coastal state or, through fishing agreements, by distant water fleets (DWF) of a developed country. Regardless of whether a developing country decides to sell its assets or develop its own fishing capacity, the position of the artisanal fishery must be safeguarded because of the economic, nutritional and employment benefits to the coastal artisanal communities. At present it is estimated that some 44% of marine fish stocks are maximally exploited and 16% are over exploited. It is unlikely that major increases in marine catches will be possible in the future, and that the most practical objective will be to maintain the present level.

Demographic trends indicate increasing pressure on the resource as more individuals seek livelihoods from fisheries and as global demand rises. Inland waters also have significant capture fisheries. They are often difficult to record, owing to their relatively diffuse nature, but they currently produce at least 10 million mt each year. With regard to rural development, generally inland fisheries can have the most impact. In floodplains, for instance, many - including women and children - engage in casual fishing which makes a significant addition to the high-grade animal protein available to a household. The extensiveness of river systems also reduces the distance over which fish need to be transported.

For example, fish from coastal areas are transported relatively long distances, with the risk of loss and spoilage before they can make a nutritional impact on the hinterland regions. Dams and reservoirs may fulfil a similar role, particularly when constructed in areas commonly short of other water bodies. Both marine and inland artisanal fisheries tend to include the poorest sectors. It is often one of the few livelihoods open to the landless and often becomes the default livelihood.

For this reason, as the human population increases and land becomes at a premium, there will be increasing pressure for people to adopt fishing as a livelihood. Aquaculture offers some response to this pressure but is far from a panacea. The commoner types of aquaculture require ownership or access to ponds and water, which often mitigates again the poorest.

The developing country decides that it does not have the resources to build its own commercial fleet but that it should sell the rights to a DWF of another nation, often of an industrial nation. Such decisions should be based on detailed knowledge of the sustainable quantities of fish to be taken since this defines the economic scope for development, but this is often not reliably known. Many countries take the easy option and enter into fishing agreements with DWF of developed countries either on a country to country basis or on a company to company basis or, in the case of the EU, on a supra-national basis.

Conflicts with artisanal fisheries may be directly for fish but may also include the price-depressing impact of fish landed locally by commercial boats. There is some evidence that more fish are being landed in developing countries now than previously. Over the last decade, for example, DWFs have accounted for around half the yield taken in the productive fisheries off West Africa. The conflict is most marked within those countries or bodies which are significant donors whilst also having major DWF fishing interests such as Japan and the EU.

The EU both negotiates fishing agreements with developing countries on behalf of the member states, whilst also having a policy of assistance to coastal states for fishery development. These responsibilities are split between two directorates until recently DG14 and DG8. In any event, adequate information is rarely available to allow sustainable levels of off-take to be determined. Information is a key factor for the sustainability of fisheries stocks. By contrast with marine fisheries, the major potential conflicts with inland fisheries come from other sectors.

Agricultural expansion is leading to a progressive modification of floodplains. More than 40% of the floodplains of Bangladesh, which themselves cover more than 69% of country, have been modified and impoldered for rice growing. The use of water resources from rivers and lakes is increasing. More than 60% of the water flow of the Ganges River is abstracted for irrigation and other purposes and whilst some is returned, the quality has suffered.

Agriculture is also increasingly using agro-chemicals which get into the water bodies where they may affect the growth and mortality of fish or accumulate in their bodies to be passed on to consumers.

Pollution in the wider sense, from industrial and domestic sources, also presents conflicts for inland fisheries. There exists, therefore, a great need for policies on inland fisheries to be closely integrated with those of agriculture, water resources and also power, where hydroelectric structures are a significant feature. Any degenerative practices in a catchment area will have a potential impact on aquatic habitats, most of all on fish and fisheries.

In fact, livelihood needs for aquaculture are far more similar to those for agriculture than to fishing itself. Indeed, farmers tend to make better fish farmers than people who primarily fish for a living. Aquaculture may have downstream impacts on other sectors where intensive culture may generate polluted effluent. When aquaculture generates high returns, this may cause land to be lost to shrimp culture at the expense of rice production.

The first thing to be clear about is that there is a distinction between policies for fisheries in development and fisheries policies as such, certainly from the point of view of donors and fishing nations. The latter will always possess an element of self-interest. The clearest example of policies is provided by the World Bank which withdrew support to the fisheries sector after a fairly fraught history, but maintains a strand of fisheries interest within the Rural Development Department.

Strong commitment to sustainable resource management is expressed in the Code of Conduct for Responsible Fisheries drawn together by FAO following the Cancun Conference in 1992. The Code takes in all aspects of fisheries and is basically a code of best practice based on existing knowledge. It includes both general principles and guidelines for all aspects of fishery and aquaculture operation and development. Article five of the Code refers specifically to the Special Requirements of Developing Countries. It indicates the need to take into account the capacity of developing countries to implement the Code, especially the poorest and small island states.

It emphasises that their needs should be supported in areas of financial and technical assistance, technology transfer, training and scientific cooperation with respect to fisheries, to enhance their abilities to develop and promote their own fisheries. There is nothing radical in the Code but it draws together preconditions for fish production to be maintained or increased into the future.

Code of Conduct for Responsible Fisheries, 1995. Non-mandatory but with obligatory elements of Law of the Sea, 1982 and Agreement to promote compliance with international conservation and management measures by fishing vessels on High Seas, 1993. The following are some key elements of FAO's Code of Conduct:

— Committed to provide assistance for Code implementations.

— Help deveoping countries to follow the Code of Conduct. States, NGOs, International organisations to promote understanding and acceptance of the Code. Assistance for regional cooperation, decision making and consultation. Control of flagged vessels and peaceful resolution of disputes to be promoted.

— Trade to be carried out in accordance with the World Trade Organisation Agreement (has no application for fish trade at present but probably will in the future).

— Use aquaculture to promote income diversification.

— Multiple use of catchment areas and coastal zone to be ensured.

— Decisions to be based on research and traditional knowledge.

— State promotes training for responsible fishing.

— Protect fish workers' rights.

— Facilities and equipment to be safe for healthy, fair work.

— Right to fish has obligation for responsibility.

— Precautionary principle is applied.

— Guidelines for protection of biodiversity.

— Guidelines for prevention of over-fishing.

— Responsibility for future of stock promoted.

Coastal and inland fisheries are mainly based on artisanal fishing with low capitalisation and mechanisation in what are essentially rural communities. Further, not all fishers are full-time, some are part-time or occasional. On the floodplains of Bangladesh only 20-30% of the total catch is taken by full-time fishers. Part-timers have other occupations, often seasonal, but fisheries remains an essential component of their livelihoods. This is a reason to recognise the interdependence of fisheries with agriculture or petty trading and, therefore, the need for a well integrated cross-sectoral development policy. Artisanal fisheries are mainly community-based and tend to be organisationally weak and have poor access to information. They are also often in debt to moneylenders since they can offer little collateral, such as land, for formal credit. This may pressurise them into over-fishing and in an open-access situation, communities and interlopers scramble competitively for dwindling resources with damaging results on the stocks.

Structured ownership or right of access by communities are crucial particularly where government institutions are weak. However, water bodies are often large and fish are very mobile so there is a limit to the extent a single community or administrative district can manage and control its own resources. There needs, therefore, to be an extensive network of linkages between participating communities to enable full management of the resources.

The impact of fishing on the stock is the most important piece of information communities need in conducting their own management. This is also typically absent from traditional management systems so far investigated. In general, mechanisms for community management are less understood in fisheries than in other sectors such as forestry or wildlife. The Code of Conduct has guidelines that are as applicable to artisanal communities as to high seas fisheries, and also highlights the institutional support required from donors. Help with information and the capacity to gather information is one of the features emphasised.

In many ways the Code points the way for future actions and fills a gap that exists in many development policies. It also underlines the fact that access to a fishery gives a responsibility for its management which underpins equally the basis of community management as it does for the role of distant water foreign fleets.

Ultimately, the hand of developing countries in international negotiations needs to be strengthened. One way of doing this is by strengthening flows of relevant information. The management of fish stocks needs regular feedback of relevant information, otherwise rational decisions on stocks under pressure cannot be taken. Most donors, particularly bilaterals, have access to considerable information collecting and analytical capacity and this is used to some extent. But the supply of information to each level is critical in assisting decision-makers.

5

Environmental Challenges

Sustainable development requires attention not just to economic growth but also to environmental and social issues. Unless the transformation of society and the management of the environment are addressed integrally along with economic growth, growth itself will be jeopardized over the longer term. Environment and social issues, when not addressed, accumulate over time and have consequences that do not show up in the shorter time horizons typical of economic policymaking.

Some study suggest that 2 billion people will be added to the world's population over the next 30 years and another billion in the following 20 years. Virtually all of this increase will be in developing countries, the bulk of it in urban areas. In these same countries, 2.5 billion to 3 billion people now live on less than $2 a day. The core challenge for development is to ensure productive work and a better quality of life for all these people. This will require substantial growth in productivity and incomes in developing countries. The challenge may seem daunting-and it is.

But over the past 30 years world population also rose by 2 billion. And this growth was accompanied by considerable progress in improving human well-being, as measured by human development indicators. Average income per capita in developing countries grew from $989 in 1980 to $1,354 in 2000. Infant mortality was cut in half, from 107 per 1,000 live births to 58, as was adult illiteracy, from 47 to 25 percent. Looking back to the 1950s and 1960s, it was feared at the time that the developing countries-particularly China, India, and Indonesia-would not be able to feed their rapidly growing populations.

In the 1960s and 1970s the Club of Rome and many other groups forecast that the Earth would rapidly run out of key natural resources. So far, this has not happened, again because changes in technology and in preferences have allowed the substitution of new resources for existing ones-for example, fiber optics in place of copper. Global action has also led to major strides in eliminating disease scourges (smallpox and river blindness), and in addressing new problems. But accompanying these achievements were some negative social and environmental patterns that must not be repeated in the next 50 years if development is to be sustained.

Today, there has been a significant drop in the percentage of people living in extreme poverty. Even the absolute number of very poor people declined between 1980 and 1998 by at least 200 million, to almost 1.2 billion in 1998. The decrease was primarily due to the decline in the number of very poor people in China as a result of its strong growth from 1980 onward. Since 1993, there have also been encouraging signs of renewed poverty reduction in India. Sub-Saharan Africa, by contrast, has seen its number of very poor people increase steadily. Yet in 1998, despite the decline in Asia and the increase in Sub-Saharan Africa, East Asia and South Asia still accounted for two-thirds of the world's very poor people, and Sub-Saharan Africa for one-quarter.

Development strategies will need to do better in eliminating abject poverty. The estimated 1 billion very poor people is of the same order of magnitude as the independently generated figures on the number of people who are undernourished and underweight. The average income in the richest 20 countries is now 37 times that in the poorest 20. This ratio has doubled in the past 40 years, mainly because of lack of growth in the poorest countries. Similar increases in inequality are found within many (but not all) countries.

In the 1990s, 46 countries were involved in conflict, primarily civil. This included more than half of the poorest countries. These conflicts have very high costs, destroying past development gains and leaving a legacy of damaged assets and mistrust that impedes future gains. The increased scale and reach of human activity have also put great pressure on local and global common property resources (water, soil, and fisheries), as well as on local and global sinks (the ability of the biosphere to absorb waste and regulate climate).

At the local level, hundreds of developing-country cities have unhealthy levels of air pollution. At the global level, the biosphere's capacity to absorb carbon dioxide without altering temperatures has been compromised because of heavy reliance on fossil fuels for energy. Global energy use traditionally has grown at the same rate as gross domestic product (GDP). Greenhouse gas (GHG) emissions will continue to grow unless a concerted effort is made to increase energy efficiency and move away from today's heavy reliance on fossil fuels.

In the past 50 years excess nitrogen-mainly from fertilizers, human sewage, and combustion of fossil fuels-has begun to overwhelm the global nitrogen cycle, giving rise to a variety of ill effects ranging from reduced soil fertility to excess nutrients in lakes, rivers, and coastal waters. On current trends, the amount of biologically available nitrogen will double in 25 years.

Fresh water consumption is rising quickly, and the availability of water in some regions is likely to become one of the most pressing issues of the 21st century. One-third of the world's people live in countries that are already experiencing moderate to high water shortages. That proportion could rise to half or more in the next 30 years unless institutions change to ensure better conservation and allocation of water. More than a billion people in low- and middle-income countries-and 50 million people in high-income countries-lacked access to safe water for drinking, personal hygiene, and domestic use in 1995.

Nearly 2 million hectares of land worldwide (23 percent of all cropland, pasture, forest, and woodland) have been degraded since the 1950s. About 39 percent of these lands are lightly degraded, 46 percent moderately degraded, and 16 percent so severely degraded that the change is too costly to reverse. Some areas face sharp losses in productivity. Grasslands do not fare much better: close to 54 percent show degradation, with 5 percent being strongly degraded. Deforestation is proceeding at a significant rate. One-fifth of all tropical forests have been cleared since 1960.

According to the Food and Agriculture Organization of the United Nations (FAO), deforestation has been concentrated in the developing world, which lost nearly 200 million hectares between 1980 and 1995. In the Brazilian Amazon annual deforestation rates varied between 11,000 and 29,000 square kilometres a year in the 1990s. Deforestation in developing countries has several causes, including the conversion of forests to large-scale ranching and plantations and the expansion of subsistence farming.

At the same time, forest cover in industrial countries is stable or even increasing slightly, although the forest ecosystem has been somewhat altered. According to a 1997 World Resources Institute (WRI) assessment, just one-fifth of the Earth's original forest remains in large, relatively natural ecosystems. Through a series of local extinctions, the ranges of many plants and animals have been reduced from those at the beginning of the century.

In addition, many plants and animals are unique to certain areas. One-third of terrestrial biodiversity, accounting for 1.4 percent of the Earth's surface, is in vulnerable "hot spots" and is threatened with complete loss in the event of natural disasters or further human encroachment. Some statistics suggest that 20 percent of all endangered species are threatened by species, introduced by human activity, alien to the locality.

The aquatic environment and its productivity are on the decline. About 58 percent of the world's coral reefs and 34 percent of all fish species are at risk from human activities. Seventy percent of the world's commercial fisheries are fully exploited or overexploited and experiencing declining yields. None of these social and environmental patterns is consistent with sustained growth in an interdependent world over the long term.

Given the social and environmental stresses caused by past development strategies, the goal of raising human well-being worldwide must be pursued through a development process that "does better"-a poverty-eliminating growth path that integrates social and environmental concerns in pursuit of the goal of sustained improvements in well-being. The development process is about change and transformation. Economies evolve.

Societies and cultures evolve at different speeds, creating stresses that need to be addressed and managed. Moreover, in an era of globalization, the growing scale and speed of change in human activity are in some cases outpacing the rate at which natural processes and life-support systems can adapt. Globalization and faster technological change are also altering the nature of social interaction and affecting the efficacy of existing institutions.

Although globalization and technological change offer many benefits, they can have deleterious side effects if institutions at local, national, and international levels do not evolve fast enough to deal with the adverse spillovers. The consequences of previous patterns of development are also beginning to bind, restricting certain growth paths or making them more costly. But these processes, if managed well, can create new opportunities.

Of the many interrelated drivers of change and transformation, four stand out: scientific and technological innovation, income growth, population growth, and urbanization. The first two are likely to continue changing preferences and providing new opportunities to satisfy these preferences.

The demographic and urban transitions, by contrast, are one-time changes, and the opportunities they offer are perhaps less well recognized. These are discussed in the next section. The flow of information and ideas, boosted greatly by the Internet, can enable developing countries to learn more rapidly from each other and from industrial countries. It can also facilitate the emergence of networks to monitor a wider array of development impacts. Other technological changes can enable developing countries to leapfrog stages in the development process that rely on inefficient uses of natural resources.

Science and technology can help address major socioeconomic problems. As noted, the green revolution was critical in enabling many developing countries to avoid widespread starvation. To benefit from these opportunities, institutions are needed that can stimulate and diffuse technological innovations and avoid or mitigate any deleterious consequences.

Increasing income growth may place a strain on the environmental and social fabric if there is too little attention to shifting consumption and production patterns. But this future economic growth will also require major investments in new human-made capital to expand capacity and to replace existing capacity as it ages. Making these investments more environmentally and socially responsible through appropriate investment criteria will go a long way toward putting development on a more sustainable path-an opportunity not to be missed.

Today's industrial countries were themselves developing, their population densities and growth rates were much lower than those of developing countries today, and the pressure on their resources was consequently lower. They also had a more evenly distributed age structure and lower dependency rates, allowing social institutions to adapt gradually to the requirements of a changing population.

Populations in industrial countries as a group were fairly stable for most of the second half of the 20th century. As a result, the growth in world population in this period has been driven primarily by population growth in developing countries. The stresses and spillovers from this population growth are generally observed not, as was originally expected, at the aggregate level but, rather, in more insidious ways-in many smaller interactions between population, poverty, and resources.

The outcomes are felt in greater pressures on fragile lands, in lower wages, and in persistent unemployment. It is now clear that a global demographic transition is well underway, even if it is not yet complete. This is a major historic opportunity. World population is expected to stabilize by the end of this century at 9 billion to 10 billion people, 20 to 30 percent lower than forecast in the 1960s and 1970s.

Many countries will have to deal sooner than expected with another problem-an aging population. This can have major consequences, especially for rural populations, for whom formal social safety nets are either nonexistent or not well developed. For example, one consequence of China's one-child policy-which dramatically and successfully lowered aggregate population-may be that by 2030 as much as one-third of the population will be over age 65. Influencing the demographic processes in many countries is the growing incidence of HIV/AIDS, malaria, and tuberculosis.

For example, current estimates and projections in Sub-Saharan Africa indicate increasingly large losses of working-age people to the AIDS epidemic. The economic impact of such high mortality is especially serious because enormous private and public investments have already been made in members of this age group. The loss of their productive lives leaves large and unpredictable gaps in the labour force. Malaria causes high levels of adult sickness rather than deaths, but this too inflicts heavy losses on labour productivity. Changes in the incidence of disease will have profound effects on health expenditures in these African countries.

With declining fertility, the age structure of the population changes, opening a window of opportunity in developing countries for a few decades-a window they can use for catching up and raising welfare for all. The proportion of the working-age population rises in relation to the proportions of children and the elderly, enabling societies to spend less on school construction and on old-age medical expenses and to invest the savings in generating economic growth. But such benefits will materialize only if the members of the working-age population are gainfully employed and have opportunities to expand their asset base.

Eventually, dependency ratios rise again as these workers age, and the window of opportunity starts to close, as it will soon begin doing in East Asia and Eastern Europe. Some regions, notably East Asia, have benefited substantially from the drop in the ratio of dependents to workers. Investment in forming a skilled, healthy labour force, combined with policy and institutional settings conducive to using this labour force effectively, helped generate strong economic growth. Two keys to success were maintenance of an open economy and investment in sectors with high growth potential.

Since most developing regions will continue to experience relatively low dependency ratios for some decades, careful preparation now can help make the most of their windows of opportunity. Until now, populations have been growing too rapidly for fiscally constrained governments to expand the provision of jobs, infrastructure, and public services enough to keep pace with people's needs. This task will become easier now that the global population is approaching stability.

Governments in both urban and rural areas can move from catching up with the quantitative need for services, to upgrading their quality. Much of the social tension and frustration arising from unemployment and poor public services can then be attenuated. Lower rates of population growth will reduce pressure on natural resources, but this will be offset by the increase in per capita consumption. The latter trend makes it essential to adopt the technologies and growth paths for production and consumption that will ensure the sustainable use of natural resources. To benefit from the opportunities a stabilizing population provides, it is critical to anticipate problems and identify development strategies for getting through the transition period without creating conditions that generate further conflict or resource degradation.

As countries move from poverty to affluence, the required growth in productivity involves a shift from heavy dependence on agriculture as a primary source of employment and income to nonagricultural activities that do not make intensive use of land. This is generally accompanied by a major shift in population from rural to urban areas. Indeed, the most important socioeconomic and cultural transformation over the past 150 years has been the transformation of relatively closed, exclusive, custom-based rural societies into relatively open, inclusive, innovation-oriented urban societies.

Rural communities, especially in less accessible areas, have long adapted to their circumstances, developing vibrant, self-sufficient communities. As long as risks could be absorbed locally, these communities continued to learn and adapt. Dependence on local ecosystems, however, imposed limits on risk taking and innovation.

This autonomous development path changes as rural areas become drawn into larger markets and strengthen their links with urban areas, making trade networks and distance from market centres more critical features of development opportunities and local resource pressures. Increasing densities in towns and cities, and the greater connectivity between cities, as well as between urban and rural areas, increases the catchment area of markets and the returns to economic endeavour. If managed well, this transformation enables the emergence of new activities and productive job opportunities.

Towns, as market centres for a rural hinterland, start the process of creating economies of scale for nonagricultural activities. Urban society also permits the spreading of risks over larger numbers of people and activities. Knowledge flows more readily, through increased opportunities for face-to-face contacts among various actors. And the need to accommodate diverse views and meet rapidly changing challenges stimulates innovation and new applications of technology. As a result, larger cities become incubators of new values-among them, risk taking and innovation. Creativity, knowledge flow, the increasing scale of activities, and larger catchment areas are central to specialization and productivity growth. This is true not just for the production of goods but also for the provision of services.

A village or neighbourhood can support a primary school or basic clinic, and the local teacher or doctor can be a generalist. But providing higher, more sophisticated, and more differentiated education and health care requires more specialized skills. Because of the fixed costs of supporting these specialized skills, a larger catchment area is required. The higher population densities, lower transport costs, and lower communications costs in towns and cities make the more specialized operations possible. In moving further up the hierarchy of required specialization, the required catchment area also increases. So, the transition from villages to towns, and from cities to metropolitan areas, corresponds to the different functional capabilities of larger, higher-density conurbations.

The potential benefits of higher densities and greater connectivity can be more easily realized if the investment climate is improved through better enabling rules and frameworks and better physical infrastructure. Stimulating and attracting investments-in particular, by the small and medium-size enterprises that provide most of the jobs for growing urban populations-is the key to accommodating the expected growth in urban populations and ensuring their ability to pay for needed urban services and amenities.

The most fundamental social and economic transformation-from traditional rural to modern urban-is manifested spatially. Except in the most populous countries, such as China and India, rural societies are relatively low in density and heavily dependent on agriculture as the primary source of employment and output.

Modern urban societies are generally higher in density and dependent on activities that benefit from proximity and do not require a great deal of land, such as manufacturing and services. These activities and land use patterns generate different types of sociocultural and environmental problems. Most ecosystems, too, are defined spatially. Much flora and fauna is locally unique and adapts gradually to changes in local circumstances.

Local problems and stresses appear earliest, whether in the form of local extinctions, the reduction of the ranges of many plants and animals, or soil, air, and water pollution. These changes, the result of local development pressures, do not show up at national and global levels until they accumulate, but they provide early warning of problematic consequences of current development patterns. The jurisdictions of many institutions that make or implement rules and laws (legislatures, constitutions, and government agencies) are also defined spatially.

Often, the spatial jurisdiction of institutions does not match the spatial nature of the social and environmental problems generated by economic activity-one reason for the persistence of these problems. People, where they live, and how they interact with each other and with nature, it is important to look at where people are now and where they are likely to be in the future. The world's population increased by more than 3.5 billion people in the past 50 years, and 85 percent of these added people were in developing and transition countries.

The number of people living in fragile rural areas in developing countries doubled, in stark contrast to the declining numbers in this category in high-income countries. The number of cities with a population of more than 10 million people went from 0 to 15 in developing countries but only from 1 to 4 in high-income countries. In the next 30 to 50 years the 2 billion to 3 billion increase in the world's population will be almost exclusively in developing and transition countries, and virtually all of it will be in urban areas. The growth of the urban population is driven by natural increase, rural-to-urban migration, and the incorporation of high-density rural areas on the urban fringe.

The number of megacities in developing countries is likely to increase to 54, while it will stabilize at 5 in high-income countries. It is not yet clear whether the number of people living in fragile areas will continue to increase, but it probably will unless migration opportunities change. As many as 2 billion people will live in two areas that are difficult to manage: fragile rural areas and megacities. Dealing with these people's needs will be a major challenge, since there is not much experience in industrial countries that can be adapted to their needs.

In Europe, 1.3 billion people living on fragile lands have modest assets that can help bring them out of extreme poverty, but these assets are seldom nurtured by local or national institutions. The people have land that is subject to many constraints, making it vulnerable to degradation, erosion, floods, and landslides. They possess human capital, which is handicapped by restrictive traditions, limited mobility, lack of voice, and poor access to services. This is even more true for women, who are thus the most marginal group. The mainly poor people on fragile lands also face circumstances vastly different from their counterparts on Europe's rural periphery 50 to 100 years ago.

Today, international migration is highly restricted, and while rural-to-urban migration is important for them, there are limited numbers of jobs at above-subsistence wages for unskilled workers, especially in the low-growth economies. As a result, as noted above, instead of declining sharply, the number of people living on fragile lands is estimated to have doubled in the past 50 years-despite some outmigration.

The problem of feeding a growing and more urban population calls for better management of the interaction with nature, particularly with respect to land and water. Whether or not rural families have land, water, and education is critical to their current livelihood, as well as to their ability to move to cities in the future. More egalitarian access to these assets is also crucial for determining the quality of society's institutions.

A successful rural-urban transition requires the elimination of poverty for those who stay in the countryside and better preparation of those who move to the cities. It also demands protection of remaining natural ecosystems and habitats, given their central role in maintaining life-support systems and biodiversity. This latter requirement is one reason to intensify agricultural production in areas already under commercial crops and pasture.

Intensification in such areas not only minimizes pressure on biodiversity and on marginal agricultural areas but also increases the food available to cities and leads to dynamic rural-urban linkages. Higher population density in these rural areas would also make investments in health and education more cost-effective and would increase the potential for off-farm employment and help farmers accept risk and innovate.

Today, cities of the developing world face a formidable undertaking, given the expected rapid rate of growth and sheer numbers of urban residents to be employed, housed, and serviced. The characteristics of periurban settlements, towns, cities, and megacities-higher density, large scale of settlement, and greater social diversity-facilitate the creation of productive employment opportunities, efficient provision of services, and access to ideas and learning. But having many people at close quarters also creates the potential for social problems-crime and social dislocation-and for environmental spillovers that pose health and safety hazards, especially for those living in neighbourhoods without sanitation or drainage and in potential disaster zones. The long life of urban physical capital stock can lock in certain development paths, making changes costly. If managed well, urban areas can be the future engines of growth.

The political, legal, and market domain for coordinating many activities is frequently the nation. Many externalities spill over beyond local communities and municipalities, and even across regional boundaries. The nation is thus often the level at which interests can be balanced, either directly or by facilitating negotiation among localities. National actors may be better placed to organize the provision of public goods and to take advantage of scale economies when the beneficiaries extend beyond subnational regions. Generating a strong investment climate, including sound macro-economic fundamentals, good governance, and basic infrastructure, requires a framework that is typically national in scope.

Dismantling perverse subsidies, husbanding forests and fisheries, and curbing water and air pollution in river basins and airsheds are major national challenges.

Managing foreign aid and avoiding civil conflict are other key national concerns that determine whether development is sustainable. At the global level. Many economic, environmental, and social processes-knowledge, conflict, disease, pollution, migration, and finance-spill over national boundaries. A few of these processes generate problems that are purely global: depletion of the stratospheric ozone layer is an example.

But most global problems and opportunities are experienced at the local level as well. Automobiles that pollute local airsheds also generate greenhouse gases; wet-land destruction that disrupts local water resources also undermines biodiversity of global significance; new ideas that are generated in one place can benefit people in other places, near and far.

The public goods nature of many of these issues and the need to address the negative externalities requires co-ordination across boundaries. The distinctive challenge for global issues is to balance interests and commit to solutions in the absence of a global authority.

Some problems of sustainability are already urgent and require immediate action; examples are local ecosystems where population is pressing on deeply degraded soils, and forests and water stocks that have been nearly depleted. In such cases productivity is already on the decline and opportunities for correction or mitigation may even have been lost; abandonment of existing practices and outmigration may be necessary. The urgency of some of these problems has been overlooked because the people most affected are physically remote from centers of power, or because their voices are not heard, or both.

Some issues call for immediate action because there are good prospects for reversing the damage to the environment at relatively low cost, as in taking measures against air and water pollution. Even then, undoing some of the damage to the affected population may not be fully possible. But knowing the health impacts does create a moral imperative to protect those affected from further exposure, to compensate them to the extent possible, and to prevent others from becoming victims. Another category of issues unfolds over a longer time horizon. The problems may not yet be urgent, but the direction of change is unmistakable. For these, it is essential to get ahead of the curve and prevent a worsening crisis before it is too costly.

Biodiversity loss and climate change are in this category: there is already a need to adapt to the consequences of past and current behaviour, but there is also still scope for mitigation, though not for complacency. Similarly, the need to anticipate urban growth by facilitating low-income settlements in safe areas and by setting aside major rights-of-way and spaces for public amenities makes it necessary to act now to avoid greater costs and regrets later. What is clear is that almost all of the challenges of sustainable development require that action be initiated in the near term, whether to confront immediate crises, such as the health risks to children from unsanitary living conditions in existing slums, or to stem the tide of crises where concerted action in the near term could avert much greater costs and disruption to human development in the longer term.

In looking back over past successes and failures in solving development problems, it is clear that there have been more successes where markets function well (for example, in providing food to people with effective demand),.even where the problems that markets have to solve (such as transport and communications) are relatively complex. The major problems that remain (inclusion, poverty reduction, deforestation, biodiversity, and global warming) are, however, generally not amenable to standard market solutions, although markets can help solve subsets of these problems.

Sustainable development is often criticized as being anti-infrastructure, as if infrastructure necessarily implied centralized, large-scale, capital-intensive, supply-side, and monopolistic facilities. Infrastructure can also mean dispersed, small or medium-scale, competitive, labour-intensive, and demand-side investments. Some experience in industrialized countries has shown that it is not infrastructure *per se* that is needed but transport, communications, water, sanitation, and energy services that can be provided in a variety of ways, on a variety of scales, and in a wide range of levels of waste, efficiency, and environmental impacts.

During the year 1990s saw the spread of the relatively new demand-based or community-based water provision approach that mobilizes community knowledge, talents, and resources to plan, design, build, operate, and maintain water supply schemes. It differs from the supply-driven or target-based approach that had line agencies bring water supply to villages as a State-provided service.

The traditional approach has long been criticized for its top-down and inflexible method and its inability to either mobilize community resources and knowledge to build better systems or to effectively maintain systems after the departure of the government agency. Tales of rural water supply schemes poorly designed, badly maintained, and inappropriately matched to community needs were transformed into calls for a new approach. The resulting demand-based approach is driven by community demonstration of interest in a water supply scheme. The government facilitates the project and provides technical assistance.

Participatory techniques determine community preferences regarding service options and design of the system. The community (i) contributes to investment costs, (ii) has significant control over how funds are managed, (iii) retains ownership of the facility, and (iv) is responsible for operation and maintenance. The approach (i) strengthens the community, (ii) lowers construction costs, (iii) reduces subsidies, (iv) delivers to households the level and type of service they are willing to pay for, (v) serves all in the community, (vi) listens to and heeds women, and (vii) keeps the facility running for its full life, with the members of the community having a strong sense of ownership. The shift of rural water supply schemes to such an approach has been uneven at best, but the process has involved considerable learning, and some successes have been achieved.

In Sri Lanka, for example, the Community Water Supply and Sanitation Project has mobilized a significant amount of community contribution: of 170 schemes analysed in one report, 163 were completed at costs below the initial estimates, with 72 completed at more than 30 percent less than projected. Communities tended to select higher levels of service than the line agency would normally have installed, and contributed more than was expected NGOs and local government units serve as intermediaries between communities and the project. They have shown high levels of competency, but have also required a large amount of training in technical matters. Although the need should have been obvious from the beginning, the initial project design did not provide for it.

Sometimes the lack of technical competence led to delays or inappropriate construction. The project has learned how to better build institutional and technical capacity. It has also grappled with the challenge of fostering more competition by increasing incentives to attract the better NGOs, the private sector, and local government units. Because demand-driven projects are new and untested, it is imperative that they be phased to test rules and processes and that they be explicitly designed for structured learning.

The Sri Lanka project was appreciably modified as its monitoring and evaluation system collected data from the field, analysed the data for successes and failures, and adjusted the rules and procedures. For example, the project changed its eligibility criteria to give priority to communities that had clearly demonstrated demand for services, the range of service levels and technical options to reflect communities' preferences for higher-than-expected levels of service, and the cost-sharing and subsidy-ceiling rules to ensure sustainability and increase service coverage. The demand-driven approach recognizes water as an economic good with a value. It also recognizes that donors and governments alone can never fully meet the demand for rural water supply construction and maintenance and that mobilization of local talents, time, and knowledge can significantly expand the pool of available resources.

The most important key to success is user participation. It significantly increases (i) overall economic benefits, (ii) the fraction of target population actually reached, and (iii) the environmental benefits of rural water supply projects. In one study, participation was qualitatively evaluated to differentiate between situations where beneficiaries were merely provided information and where they had a decision-making role, and to identify whether beneficiaries had a role in design, construction, operation, and maintenance. In general, "in communities where household members made informed choices about whether to build a system, the type of system to be built, and the level of service to be provided, sustainability of water services was higher".

Preexisting health education, community organization for cooperative ventures, mechanical skill development, accountability procedures for government line agency personnel, and experience in decision making with respect to community infrastructure can greatly increase the likelihood of success in a water supply project. Success in a water supply project can also lay the foundation for success in other community mobilization and cooperation ventures.

In Philippines, UV Waterworks technology involves the use of high-dose ultraviolet (UV) light and gravity-flow design to kill almost all bacteria, viruses, and Cryptosporidium in drinking water, at an extremely low cost. The amount of UV light used to disinfect water of pathogens is more than three times the US Environmental Protection Agency standard. With the use of the technology, cleaning 1,000 l of water costs less than \$0.05.

The UV Waterworks device has no moving parts and relies on gravity for water flow. Minimal electric power is used only for the UV light, and power from car batteries and generators, and solar, wind, or hydropower can also be tapped. The technology is employed by WaterHealth International Inc., which distributes UV Waterworks in various countries.

As early as 1997, Bendix Sales Corporation (BSC) of the Philippines was already planning to open water stores using the UV Waterworks technology. A UV Waterworks water store offers a profitable retail opportunity for entrepreneurs and provides low-cost water in poor communities. Each water store has the capacity to deliver as many as 20,800 l or 5,000 gal of water per day. BSC started selling UV Waterworks-treated water in 1998. By May 1999, BSC had opened 37 Aqua Sure water stations using the UV Waterworks technology in urban areas and 30 community water centres in rural and urban poor areas such as Smokey Mountain in Tondo, Manila. In such areas, people without access to reliable drinking water can buy UV Waterworks-treated water at one third the cost of bottled water.

The availability of UV Waterworks-treated water has helped eliminate the practice of boiling water using firewood and in reducing waterborne diseases in both urban and rural areas. To date, approximately 50,000 people are being served by UV Waterworks-treated water stations in the Philippines. Other water products using the UV Waterworks technology have also been made available for distribution in medical missions, public schools, and public health clinics. The successful operations of BSC's urban water stations and rural community water centres demonstrate that UV Waterworks technology is an effective and financially viable method of providing affordable clean water to people, including poor communities.

In Asia and the Pacific, municipal water monopolies whether public or private, struggle to keep up with the water demand of the industrial, commercial, and middle-class residential sectors. They generally have (i) large unaccounted-for water losses; (ii) tariffs far below economic (considering (a) the economies of long-run marginal cost of supply, (b) opportunity cost of water in alternative uses, and (c) externalities of water supply development); (iii) serious supply and distribution problems; and (iv) no independent government regulation. They have no formal and legal competition, and financial costs and risks are usually borne by the government. Experts advise privatization, which is no panacea.

Privatization has been done in the absence of regulation, with governments bearing the financial risks, and in the absence of steps to increase competition in supply or distribution, with all attention to the supply side and none to demand-side management. It ignores environmental externalities and has proceeded with no public participation.

International water cartels have benefited and efficiency marginally improved in a few instances, but serious financial, coverage, and environmental problems remain. Many Asian cities have a flourishing water supply and sanitation private sector that serves the poor, unconnected areas of the city. It is composed of a large number of diverse actors, from households with connections selling water to suburban truckers with their own tube wells, NGOs organizing water sales at municipal water points or installing public latrines and bathhouses, night-soil carriers in India and the prc, and even a few private drinking water and wastewater treatment plants. The urban poor commonly purchase water from vendors and pay unit costs 20-100 times the meter tariff rate paid by the rich.

This private sector, invisible to consultants and civil servants, must recover full cost or it is out of business. It has essentially no unaccounted- for water and receives no public subsidy. Only a few NGO schemes are not financially sustainable. Unlike water monopolies, small players are responsive to demand. They charge market prices that cover costs, and understand willingness to pay, assume all financial risks, reach the poor, maintain close customer relations, often provide varied related services, rapidly respond to changes in demand, try out and market-test new technologies, and disseminate new technologies quickly once they recognize market demand. They flourish under competition and in the absence of misdirected subsidies. Water provision by small players may be only an interim solution, but its replacement by cheap, high-quality piped water directly into every home is still a distant possibility.

Unregulated private extraction of groundwater can have serious detrimental environmental effects. In Manila, for example, private wells have accelerated saltwater intrusion into aquifers. In areas such as southern India, water tables have dropped precipitously. Environmental damage is not caused by small-scale water entrepreneurs alone, but it is a generic issue in water resource management involving (i) allocation of water among sectors, (ii) environmental externalities of water supply and wastewater disposal, (iii) legal rights to water, and (iv) regulation of all phases of the water sector. Unserved areas in cities are disproportionately poor, and without the State's strong social or contractual commitment to serve them, attempts at cross subsidization through lifeline rates or rising block tariffs can discourage expansion by a profit-driven utility, whether public or private. Connection fees may be too expensive for poor households, and pro-poor policies should include modest subsidies of such fees or a system of several monthly payments.

Financing of expansion costs should also be spread over all users of the service since most of them have paid little or nothing for expansion to their area. In-kind payments, primarily labour, can also be mobilized as a substitute for cash payments for connections. Community participation in planning and design of water supply and sanitation, including decision making regarding the level of service to be provided, can be a valuable way to reconcile costs with willingness to pay and affordability.

In Bangladesh, approximately 70 percent of Dhaka's 2 million squatter dwellers are deprived of safe drinking water because they lack titles to the land they occupy. In very poor areas, the solution had been the provision of free street hydrants, yet it did not allow cost recovery. With the support of local NGOS and international donor assistance, communities are being allowed to manage water points. The community bears the cost of installation and pays a security deposit for water points on city property. The community is tasked to provide access roads and do excavation while the water authority lays pipes, installs the water points, and supplies water. This has resulted in access to permanent and legal water sources at half the price. Although prices are low, cost recovery has been possible.

In some places such as Manila they stress low water rates and ignore water efficiency, water reuse, or demand-side management. Metering, however, is encouraged or mandated in some cases. The scarcity value of water is not included in many cost structures; in Manila, for example, the concessionaries pay nothing for bulk water supply from Angat Dam. They do not explicitly aim to reach the poor, although some do have mandatory expansion clauses. Large municipal water supply and sanitation companies are sometimes overwhelmed by the difficulties of providing high-quality service. As public monopolies, they often do not employ least-cost production methods, are not innovative, and often fail to provide the quantity and quality of service for which customers are willing and able to pay.

Privatization has not been preceded by the establishment of a regulatory structure. Many argue that privatization can increase funds available for the sector's enormous investment needs. While it is undoubtedly true with respect to the profitable parts of the municipal water supply business, if the public sector is unwilling or unable to invest in water supply or sewerage systems for the poor, then there is no reason to believe that the private sector would be any more willing. Decentralization may open up new possibilities for urban community groups in slums to work with local government units. However, as indicated earlier, decentralization in the absence of local governance capacity building could be counterproductive.

Traditional water supply agencies or corporations and ngos can, with support, work together to develop community approaches to water supply and sanitation. In India, for example, microfinance is in great demand and has been used to improve housing, roads, electricity connections, and water supply and sanitation. Microfinance is an important enabling tool but must be supplemented with community organization building, institutional support to link up with the local government unit, improvements in land security status, and woman-focused programs in skills development and financial management.

Today, Asia Facing one of the greatest environmental challenges is to improve sanitation in both urban and rural areas. Existing levels of coverage and quality are low and not improving rapidly. Both the number and the proportion of people without access to sanitation continue to increase despite the expenditure of large sums on bringing sanitation to urban dwellers. Even where sewerage systems serve a large fraction of a city's population, treatment of the collected wastes is generally inadequate or nonexistent. And also face widespread water pollution and acute threats to human health.

In cities such as Jakarta and Manila, surface water and groundwater are often contaminated and unusable for drinking because of discharge of untreated wastewater either from sewerage systems or from individual cesspools that drain directly into watercourses. Seafood contamination by raw sewage has been responsible for hepatitis outbreaks in the prc and other countries. The supply-driven, top-down, centralized approach has failed. Policy failure, insufficient government commitment, and inadequate financial resources have undoubtedly played a role in its failure. These are often compounded by and associated with lack of governance, and consumption has made the problem very serious.

While the elite has escaped the worst of the mess, the disjunction between government capability and the dismal reality of life in an urban slum has become overwhelming. Master planning "modern" sewerage systems in megacities is, for the most part, an impossible challenge. The approach that holds most promise of both dramatically improving the urban environment and reaching poor neighbourhoods is demand driven and based upon mobilization of individual, neighbourhood, local government, and line agency resources. It harnesses a synergistic combination of government, private sector, and civil-society talents in a cooperative and competitive environment. It selects technologies of a scale and service quality strongly influenced people's expressed preferences and willingness to pay. It also mobilizes resources at the community level and creates the political pressure to mobilize commitment and resources at the local and national government levels. Technology, however, is not the principal issue.

Political mobilization, public participation, community institutional development, and the process of cleaning up the urban environment are the arenas of most leverage and action. The following are the essential components of an approach to urban sanitation:

— User preferences determine services to be provided.

— Sanitation services are unbundled into discrete vertical (household, feeder, trunk, treatment, reuse) or horizontal (geographic) units.

— NGOs and people's organizations intermediate among households, communities, local government units, and government sewerage authorities.

— Innovative financing and cross-subsidization schemes based on unbundling make sanitation affordable to the poor.

— A range of private and public sector players are involved in co-production of sanitation services at different levels (household, community, feeder, trunk, treatment) and in fostering competition and corporate competence.

— Households and communities share the cost of sanitation improvement. Men and women both undergo hygiene education and environmental awareness training.

— Community organizations, ngos, local government units, and private sector enterprises undergo technical, management, financial, and organizational skills development necessary to provide effective sanitation services.

In Pakistan, the Orangi Pilot Project has become a famous example of people empowerment. Initially, the project focused on developing low-cost ways of granting the people's strongly expressed wish to rid their environment of excreta and wastewater. With the help of committed project architects and engineers, and the residents' own labour, the communities installed in-house sanitary latrines, house drains, and shallow sewers in the lanes and streets at an average cost of about $25 per household.

Householders contribute their share of the costs, participate in construction, and elect a "lane manager" to represent about 20-30 households. These impressive achievements required the overcoming of four barriers: i) the psychological barrier posed by peoples' belief that developing a sewerage system is the duty of government and should be given free of cost; ii) the economic barrier posed by the high cost of the conventional system designed to specifications developed in rich countries; iii) the technical barrier requiring the training of local residents and their contractors in the skills necessary to construct an underground sanitation system; and iv) the sociological barrier requiring organizational skills for community collective action with full public participation.

The Orangi Project still stands as a model of community development and community sanitation improvement. It has shown that poor urban households are willing to pay lease and development charges to a municipal authority if they can overcome their mistrust of government projects and agencies. Slum-upgrading projects can become nearly self-financing after the initial grants to support community organizing and skills training. Trunk sewers and treatment plants, however, must be financed by government and donors. The government should coordinate projects while ngos catalyse and facilitate them. Once government line agencies or local government units accept ownership of the project, the key becomes the people's organizations themselves.

People's organizations need to (i) define the project in terms that are right for them, (ii) accept major financial and organizational responsibility, and (iii) keep up the pressure on government to uphold its part of the bargain. There are many models for community-State cooperation, and generalized prescriptions are not possible. It is ironic that much of the solid waste management in Asian cities is accomplished by the poor and more effectively than the public or private contractor sector, with greater resource recovery at far less cost, and with signal lack of recognition.

The urban environmental movement is predominantly middle class and strongly favours slum clearance and removal of disagreeable activities (landfills, for example) to someone else's locality. The mayor who is firm in the resolve to demolish slums and squatter settlements is often very popular with the media, urban elite, and the middle classes. The high level of cooperation among low-income residents striving to build a decent life in the face of a hostile market and State is most striking. It makes abundantly clear how sterile the economists' notion is of the individual as an autonomous decision maker participating in competitive market exchange and driving economic development.

Cooperation, trust, and mutual obligation are the foundations of any economy and the true building blocks of sustainable community development. Unfortunately, State policies often destroy community solidarity by dividing, relocating, and intentionally splitting poor communities through arbitrary land security and infrastructure policies, and by subjecting them to violence and intimidation.

In India, the Sulabh International Social Service Organization is an NGO working to promote sanitation and prevent environmental pollution in India. It is one of the largest NGOs in the world, with over 50,000 volunteers. Sanitation, a basic condition for development, is very poor in India. It is reported that only 10 percent of rural areas have sanitary facilities.

Sanitation-related diseases remain a major cause of death in the country. At high risk are half a million scavengers in India who collect and carry human excreta. Sulabh International has worked to develop a cheap and effective sanitary toilet-a two-pit, pour-flush, water-seal system that does not need sewers or expensive treatment plants.

The technology abolished the need for manual handling of human waste, thus improving environmental sanitation, community health, and community hygiene. It is also a cost-effective and appropriate technology requiring only a fifth of the water used by conventional water closets. Sulabh International has constructed and converted over 1 million household toilets and over 3,100 public toilets. Over 37,500 scavengers have stopped manual collection and 3,500 of their dependents have been provided with vocational training. There are now 24 scavenger-free towns in India. Sulabh International is an example of how NGOs can have an important impact on human development. Through the development of appropriate technology and its effective application, NGOs can have significant impacts on people's quality of life.

Madras-based Civic Exnora-Waste Management Systems were conceptualized by a NGO-Exnora International, which promotes the formation of neighbourhood associations for managing waste collection. A Civic Exnora is formed by households from one or more streets and is managed by a committee of elected or volunteer office bearers. Using bank loans or funds from a private sponsor, a tricycle waste collection cart is operated by appointed and trained waste collectors.

Civic Exnoras are primary waste collection systems that function only in areas with secondary waste collection systems or sufficiently large, open dumping spaces. This is the main reason why the success of Civic Exnoras has not been experienced in many other suburbs of Madras. Its uniqueness lies in its recognition of the lack of an institutional structure to represent the interests of the local people. It is also a venue for getting the attention of civic authorities to redress the problem, particularly in Madras, which for 25 years did not have a local election.

In Southwestern Bangkok Chonglom is a slum of about 180 households. Most of the houses are owned by their approximately 900 residents who are engaged in a wide range of economic activities. Until a few years ago, all residents experienced severely degraded environmental conditions. Lanes between houses and shops were makeshift constructions of wooden planks that ran above the often flooded, usually smelly, trash-ridden land below.

Houses were densely packed into the area, which had few common facilities. Although piped water and electrical lines had been installed after direct elections of parliamentarians representing the area began in the 1980s, the community still had the ambiance of an unhealthy slum.

By the early 1990s, conditions had dramatically changed for the better. The turnaround began in the late 1980s when the community was approached by two Mahidol University faculty members who had arranged the possibility of a $50,000 interest-free loan from Citibank for community improvement. The stipulation made by the Mahidol professors was that, in addition to setting loan repayment schedules, the community must organize itself and set its own priorities.

After some reluctance due to suspicion of the outsiders' intentions and problems among rival community leaders, the community took the decision to accept a loan, with the first project being a massive cleanup and construction of safe cement lanes throughout the community. Households and hired labour joined in the project. Soon all 624 m of lanes in the community were paved. With lanes safe to walk in and the ambiance decidedly improved, small shops started springing up in the community. Beauty shops, video rental stores, food stalls, and mini dry-goods stores appeared in the front of homes along the pathways. Many were owned and run by women. The success of the community's first efforts led to a long string of other projects. Among the major ones were piped-water connections to all houses and, in parts of the community, connection of a new drainage system with the city's.

In 1990, the community was selected as the best in all of Bangkok. Having discovered that within the space of less than five years Wat Chonglom had made itself into a decidedly livable community, the State began to embrace self-help efforts as an archetype for the rest of the city. The National Economic and Social Development Board, the national planning body, adopted Wat Chonglom's "model" of community empowerment. Another initiative was the inauguration of the Urban Community Development Office (UCDO) in 1992. Established by the Government with a $50-million fund, it is designed to be a "nongovernment government agency" that, although government-sponsored and under the National Housing Authority, is "semiautonomous" in its decision making.

Democratization processes in many countries, strong traditions of social responsibility in transition economies, and a plethora of dynamic and capable ngos and community organizations give reasons for optimism. Communities need access to public resources and support. They do not necessarily need infrastructure subsidies, but they do need connections to piped potable water, electricity, roads, telecommunications, and either trunk sewers or a sewage system. They also need technical and organizational support. Their continued development is dependent upon education, basic literacy, skills development, and health services, especially preventive health and sanitation, nutrition education, and immunizations.

Thailand has a very low level of environmental infrastructure and services given its moderately high per capita income. Only 2 percent of Bangkok's population is connected to sewers. But Wat Chonglom is an example of how a slum can be dramatically improved through a combination of catalysts from NGOs and academia, a modest loan fund provided by a large corporation, and a community with the following:

— a modicum of tenure security;

— households relatively stable and large enough to allow division of labour to support community activities;

— already-existing inter-household networks; and

— trusted and effective leadership.

Suppliers of telecommunications services are usually monopolies that have few incentives to commit sufficient resources to make rural operations a success. Remote regions are the least attractive. Rural areas thus have poor levels of service and are considered to represent a financial loss. The reality, however, is that business, institutional, and payphone use in rural areas can now produce high revenues at low cost and yield high socioeconomic benefits to a broad spectrum of the community. The development of the rural nonfarm economy requires telecommunications services. The services do not directly benefit the poor and should not be justified on poverty-reduction grounds.

New technologies such as Mobile Personal Communication Systems by Satellite are rapidly emerging and making lower-cost rural hook-ups more feasible. Countries in the region may thus surpass the industrialized countries in providing rural access, although not in the level of penetration of the service.

Thailand is aiming to have at least several payphones in each village and calculates that the plan is both financially feasible and economically very attractive. Thailand is encouraging build-operate- transfer competition while other countries such as the Philippines are issuing licenses subject to the provision of certain specified levels of rural services.

Telecommunications can aid the exchange of information crucial to the construction of sustainable rural societies, including extension advice on organic agriculture from the internet, instantaneous market information, updates on environmental information such as the hazards of particular pesticides, distance learning to upgrade the quality of rural schools, instant access to expert medical advice, and experiences of people's organizations all over the world. It is not clear, however, that the poor derive any special advantage from the widely touted internet revolution, which has been a revolution by and for the rich. Transport infrastructure can be environmentally and socially damaging. Road alignments and construction in hilly or mountainous areas cause severe erosion and massive soil loss. The opening of motorized access to forest areas can have dramatic negative impacts on biodiversity through logging, hunting, and land clearing.

Highway, rail, port, or airport construction can seriously affect wetlands. Increases in air, noise, and water pollution from vehicular movements can be severe. Dislocation of human settlements is nearly always major and disproportionately borne by the poor. It is no wonder that transport infrastructure is contentious. These impacts are exacerbated by the strong urban, male, and nonpoor bias of most transport sector investments. They are justified as being necessary for general economic development. The justification would be sufficient if it were based on a thorough examination of all the options and detrimental externalities involved in construction and utilization of the transport infrastructure being evaluated.

Unfortunately, comprehensive assessments are rarely, if ever, performed because of the narrow mind-sets and preconceived notions of transport planners, who cannot see through the eyes of the poor, women, or a forest ecosystem. Transport systems are currently a major source of unsustainability in the industrial system. They are economical only as long as their externalities can be fobbed off on the voiceless. They depend upon the supply of cheap liquid fuels; impose exorbitant health costs on urban neighbourhoods; maim, injure, and kill millions of people a year in Asia and the Pacific; create strategic dependencies on imported fuels from unstable regions of the world; affect the sleep, concentration, and intelligence of every urban and many a rural resident through noise and air pollution; and get away with all the devastation through enormous, corrupting political and market power.

Table: 1Use of Solid Fuels in Asia and the Pacific (millions of users)

	Biomass	Coal
People's Republic of China	630	370
India	650	20
Southeast Asia and the Pacific	430	10

The transport sector is the most subsidized and centrally planned sector in most of the world's economies. The expense and scale of transport infrastructure projects crowd out hundreds if not thousands of smaller, human-scale projects whose total economic impacts may be far greater. Transport infrastructure serves well the needs of only a fraction of the citizens of the region. It is an anti-poor and anti-environment anachronism badly in need of competition, innovation, and holistic redesign.

Telecommunications facilitates the performance of other economic undertakings and is crucial to the performance of information- and knowledge-based activities.

But in rural areas, telecommunications services can not only influence production and other economic decisions of rural households, but can also do much to address the information asymmetries that are one of the important causes of inequality and poverty. The services can have dramatic positive social impacts.

A sustainable transport infrastructure strategy would seek to encourage the development of urban areas so that infrastructure needs are met in a cost-effective manner while enhancing human habitability. Urban development must free itself from the speculation, rent seeking, corruption, collusion, fiscal disaster, environmental catastrophe, and anarchy that is urban development today.

The rapid growth of Asian economies and urban centers-with its large amounts of new investment in urban construction and expansion-means that it is still possible to adjust land-use patterns to achieve a better balance between public and private transport and between motorized and nonmotorized modes. Cities have many regulatory and fiscal options to influence land-use patterns so that revenues are raised for improving urban services to all areas of the city, rents from development stimulated by infrastructure investments are captured by public bodies for the public good, and the community is built rather than commuting amplified.

Table: 2 Indoor Air Pollutants from Typical Asian Wood Stoves

Component	Mechanisms of health effects	WHO standards mg/m 3	Wood stove mg/m 3
	Carbon monoxide Inhalation into respira tory system	10	150
	Absorption into blood from lungs Elevated carboxyhye- moglobin (HbCO) levels Reduced oxygen to body tissues Possible cilia-static impact on lung clearance		
	Particulates Inhalation into respiratory system	0.1	3.3
	Deposition in respiratory tract Irritation and toxicity		
	Benzene Benzo(a)pyrene: Inhalation into respiratory system	0.002	0.8
	Deposition and absorption in lungs Metabolic activation Precursor to cancer		
	Formaldehyde Irritation of mucosa	0.1	0.7
	Toxicity to cilia Reduction in lung clearance ability Possible carcinogen		

A pro-poor and gender-sensitive transport policy would focus on rural roads, bridges, and trails; intermediate modes of transport such as carts, rikshas, bicycles, and scooters or motorbikes; urban walkways and bikeways; and rail, especially in high-density areas. Careful coordination of employment-guarantee schemes and transport infrastructure needs can lead to more cost-effective construction and maintenance of people-oriented investments. Bicycles are friendly to the poor and to the environment. Cyclists consume 19 calories/km compared to automobiles' 1,080. Cyclists' respiration contributes a bit more than 1 g of carbon dioxide per passenger kilometre; a single-passenger automobile contributes 51. On the road, 8 bikes occupy the space of one car; 20 bikes can be parked in the space needed by one car.

The poorest women and children in many rural communities spend a great deal of time travelling. They fetch water and fuel, lug agricultural produce from field to home, and go back and forth to market. Children are often pulled out of school to help. Women are often discouraged by cultural norms and fear of sexual harassment to use public transport. Women's multiple tasks mean many trips, often during off-peak hours and over less-used routes. In many countries, women are always at a disadvantage when it comes to travelling to health clinics, schools, government offices, and markets. Improved transport can increase women's productivity, open up economic opportunities for women, and promote equity. Yet, for the most part, these benefits are, at best, afterthoughts in transport planning, tidbits thrown the way of gender goals.

Public participation that actually looked at the needs of women and the socially excluded might come up with radically different transport systems. The arguments for market intervention in support of infrastructure services provided by small and medium-sized enterprises in the public, private, and non-profit sectors are many and diverse. The most important are the following:

— The number of mostly poor unserved and underserved households and small businesses is in the hundreds of millions to billions, depending on the particular infrastructure service.

— Infrastructure services are necessary (if not sufficient) for poverty alleviation and broad, equitable economic development.

— Many of the services can be provided on a commercially viable and affordable basis. Where the services are provided in a financially sustainable manner, it is often by small and medium-sized enterprises.

— The providers (or potential providers) face market and policy constraints such as lack of access to financing, need to strengthen business capabilities, and regulatory barriers to entry and/or expansion.

— In addition to the direct benefits resulting from service provision, small and medium-sized enterprises often produce employment and other economic benefits because of their linkages with other sectors and multiplier effects.

Rural people without access to telecommunications services and/or electricity or other infrastructure in Asia and the Pacific number in the many hundreds of millions. To this must be added the often-large number of unserved and underserved in urban areas. They represent an enormous potential market. It is usually the poor or near-poor and their communities (urban and rural) that are unserved or underserved. The underserved pay more-often much more-for their lower levels of service than others. In contrast with the well-off, the poor frequently pay more both through direct out-of-pocket costs for kerosene, for example, or for dry-cell batteries for lighting, and through the opportunity costs of having to carry water from a distant source or using scarce fuels to boil drinking water. While the evidence of the poor paying more is perhaps more widely documented in the case of potable water services, it has also been well documented with regard to energy and telecommunications services.

The following are the degrees of financial sustainability:

— *Maximum sustainability*: Prices the small and medium-sized enterprises charge for the infrastructure services they provide cover all hard and soft costs plus the enterprises' target return on equity.

— *Conventional sustainability*: Similar to maximum sustainability, except that it does not cover some or all of the soft costs.

— *Partial sustainability*: Lower than conventional sustainability. Prices cover all operation, maintenance, and repair costs, no soft costs, and only some hard costs (capital and return on equity).

— *Operation, maintenance, and repair sustainability*: Prices cover only operation, maintenance, and repair costs.

— *Nonsustainability*: Prices do not even cover operation, maintenance, or repair costs.

Commercial sustainability implies the investment earns the investor's target financial return on equity-the return being after-debt service and payment of operating expenses as well as allowances for recovery or replacement of capital investments. These definitions of sustainability do not preclude implicit or explicit financial subsidies that may come into play via a public-private collaboration or through the provision of development finance or development assistance.

With vehicular traffic accounting for 90 percent of all atmospheric lead emissions in Bangkok, the National Energy Policy Office (NEPO) developed a strategy to introduce unleaded gasoline and phase out leaded gasoline.

NEPO used a collaborative approach of targeting stakeholders, identifying decision makers, holding discussions, and building consensus among State-owned petroleum companies, multinational oil companies, domestic oil companies, government agencies, and motor vehicle producers. Targets were set for the voluntary sale of unleaded gasoline, reduction of lead content in gasoline, equipping new cars with catalytic converters, and the complete phaseout of leaded gasoline.

The Nepal Government with support from the Dutch development agency is implementing a successful Biogas Support Program (BSP) that has greatly increased the penetration of household biogas plants such that there are now over 50,000 in the country. Important parts of the program have been strict quality controls, equipment warranties, an explicit biogas capital cost subsidy, and utilization of an existing and implicitly subsidized loan program. One outcome has been the emergence of a growing biogas supplier industry, although the full competitiveness of the market place has yet to be established. Despite its ongoing success, even greater and equitable penetration of the biogas program is constrained by the need for loan collateral of the type acceptable to the banks.

Many households do not have such collateral even though they have the animal and dung resources, water access, and loan payment ability needed to invest in biogas plants under the program. A pilot loan guarantee option was set up and organized to provide the needed collateral and to under- take the required financial, social/organizational, and technical intermediation. A guarantee fund under the auspices of a local activist NGO was set up at a local bank and used as collateral by the bank for making loans for biogas plants to households organized by the NGO.

Today, commercial energy is changing, not from choice but from necessity. The end of the era of cheap oil is in sight, the environmental externalities of coal use are too great to continue to ignore, and a technology revolution has undermined arguments for centralization of electric power-generating facilities. In some cases, the changes are being enhanced by regulatory moves to (i) restructure electric power sectors, (ii) internalize at least some of the environmental costs of power generation, and (iii) allow a modicum of competition on a more level playing field. Competition springs from a wide variety of sources. It is unclear where the changes are heading, but what is certain is that the days of gargantuan power stations, centralized utility grids, and energy monopolies are numbered.

The dynamic changes that will transform much of Asia in the next several decades offer the opportunity to correct the errors committed in industrialized nations. Since the early 1990s, for example, developing countries have allowed and encouraged the private sector to enter into build-operate-transfer and build-own- operate contracts with State-owned utilities for large-scale electricity supply expansion. Partial information indicates that a total developing country capacity in excess of 48,000 MW is being financed this way, with a further 24,000 MW in planning or negotiation stages.

Independent power producers in Indonesia, Malaysia, and the Philippines already generate a significant proportion of total power. Independent power production has become big business with influential interests supporting its expansion. The process, however, contributed to the Asian financial crisis and has left some national utilities in severely strained circumstances as it has built long-term rigidities into electricity systems.

The energy business is poised to become a service industry, operating with flexibility in a decentralized milieu and responding to the diverse service needs of households and enterprises. It will be characterized by competition and entrepreneurial risk taking. It needs government help in the form of (i) the legal and regulatory environments to flourish, (ii) transparency and accountability to enhance true competition, and (iii) access to the information required to make sound business decisions.

For example, rather than subsidize a centralized rural electrical grid extension motivated by military security and political considerations as much as by economics, the State could encourage a decentralized business environment that would spur a wide range of initiatives by private and non-profit sectors to deliver needed and affordable energy services such as better cooking, lighting, refrigeration, telecommunication, and motive power. The State would not eschew subsidies but would carefully craft them to support health, environmental, equity, and poverty-reduction efforts.

During the year 1991, Thailand's petroleum product sector had six oil companies- two public and four private, with the latter dominated by large international companies. Both prices and market entry were regulated, but there were various forms of non-price competition. Thailand's economic growth saw a dramatic increase in motor vehicle ownership and use and, especially in Bangkok, serious air pollution. The use of leaded gasoline and Bangkok's land-use patterns resulted in very high lead exposure for the population. Blood lead levels of Bangkok residents were among the highest in the world.

A 1989-1990 study of infants found lead levels nearly twice the levels that the US Centre for Disease Control and Prevention considers "dangerous." Deregulation began in 1991. Since then, the refining and marketing business has become much more competitive. There are now 15 oil companies in the market, including aggressive local independent companies. Coinciding with the early years of deregulation was a program to phase out leaded gasoline.

Thailand's petrolcum product consumers have benefited from both price competition and increased convenience associated with wider presence of retail stations. All residents have benefited from the elimination of lead-derived pollutants. The changes were brought about by multidimensional policies that involved many players on the Government side and public-private approaches to reach consensus on timing and other factors. Old-fashioned styles and operating modes of monopolistic electric utilities will have to change if a competitive and decentralized energy service business is to develop. Vested interests are as powerful in power infrastructure as in municipal water supply and sanitation, transport, and telecommunications.

The existing system has incentive structures that promote inefficiency, resistance to change, and large "nontechnical" losses. The issue is not ownership, but management or corporate governance. Encouraging participation of small and medium-sized enterprises in the energy service business will facilitate new technological adaptations such as integration of electrical and telecommunications systems. Also likely is the creation of "virtual" utilities that provide household or institutional electricity on a fee-for-service basis but through dispersed provision of locally tailored power and demand-matching appliances.

A wide range of energy service companies and public-private partnerships could develop on a custom basis (with services tailored to local resources and needs), with the State providing information access, creative financial (and sometimes social) intermediation, and policy changes. The developing countries of Asia and the Pacific desperately need to increase useful energy at end use. They are grossly inefficient in energy use, whether it be energy expended collecting fuel and water, cooking, or producing iron and steel.

Building factories to produce efficient motors, boilers, appliances, lamps, and windows, or building heating, ventilation, and air conditioning systems will require much less capital per unit of extra comfort or light or other service, yet will be more labour intensive than a central power station and its associated transmission lines.

Information and management intensities constitute a hidden barrier due to the scarcity and, hence, high opportunity cost of many industrial managers' time and the time required to adequately investigate sustainable options. The situation is exacerbated by the absence of players who specialize in providing the service. Only if energy represents a large portion of production costs and/or market conditions are particularly competitive may managers have sufficient incentives to examine more sustainable options such as cogeneration.

All private power projects have considerable opportunity costs for preinvestment studies, negotiations, and permitting. These costs can be a barrier for private developers because the preinvestment information intensity combined with other preinvestment costs can be as great or greater than larger conventional projects, so the developer often prefers to investigate larger projects. Another factor is the difference between the opportunity cost of capital (or, alternatively, discount rate) of industrial, household, or other potential investors in site-specific sustainable investments (e.g., in energy and materials efficiency improvement, renewable-energy-based heat and/or electricity production) and the lower cost of capital of investors in conventional options-utilities, oil companies, and the like.

The successful market penetration of any investment embodying a technically feasible technological option that is "in principle" commercially viable is inevitably dependent on the organization and chain of enterprises providing supporting or complementary inputs or services (including financing, and operation and maintenance). Frequently, the adage "the chain is only as strong as its weakest link" holds. The capability of the chain of technical and financial intermediation is critical for market penetration of small and medium-sized alternatives.

Governments and development finance institutions should ensure that renewable-energy-based options (and hybrids of renewable and conventional options) are given due consideration in rural electrification and other rural energy programs. Notwithstanding their difficulties in covering operating costs, many developing-country utilities are often required to extend service under rural electrification programs, sometimes at the same price at which they provide service in urban areas. Given the remoteness of many communities from the existing grid, their typically low load factors, and frequent collection difficulties, the grid extensions can drain revenue from the utility.

In contrast, entrepreneur-, community-, and/or cooperative-operated mini-grids based on renewable energy, and household systems based on battery charging and photovoltaics are not only potentially financially viable but also have remarkably good collection records. Local entrepreneurs and organizations-local government units, ngos, and community-based organizations, for example-also mobilize rural economic development.

The development of small-scale energy service companies can expand the absorptive/ adaptive capacity of the local private sector and of local capital markets. It can involve the following:

- local energy resource utilization by local government units, cooperatives, and community-based organizations;
- local non-energy-resource mobilization,

— local capital market development, because, with appropriate financing instruments and approaches, much of the financing is within the absorptive capacity of local capital markets;

— local industry (construction, metal, mechanical, electrical, and other) linkages and multiplier effects due to the very high local content that is possible, for both wholesale- and retail-type small-scale energy investments.

Rural small-scale energy investments and rural development should mean more equitable growth-that is, betterment for all but with particular concern for the poor and the socially excluded. The development process and poverty alleviation must be accompanied by energy system transformations that allow the introduction and/or increased use of modern energy conversion devices to decrease human drudgery, increase productivity, increase the quality of life, and facilitate information transfer in order to reinforce the other changes.

To achieve the desired development linkages with rural small-scale energy investments, it is often critical that the scope of investment go beyond primary energy development and encompass "downstream" energy technologies. Such investment will then produce the requisite energy transformations and deliver the forms of energy necessary to facilitate rural development by increasing productivity, reducing drudgery, and improving the quality of rural life.

It is the energy-transformation services such as milling, grinding, pumping, and lifting that fuel development. Their mere availability may not be sufficient for rural development but it is absolutely necessary. Achieving economic development requires attention to how investments are implemented and to the equity dimension of rural energy investments, particularly concerning women and children.

Specifically, women should be involved as trainees, entrepreneurs, decision makers, operators, builders, and owners of rural energy service small and medium-sized enterprises. If small-scale energy investments are to penetrate a market and be commercially viable, they must have access to both equipment and technology.

Technical intermediation services must be available in both rural and urban markets. Equally necessary are financial intermediation services that provide accessible and appropriate medium- and long-term financing in both urban and rural markets.

The organizational chain has to perform in two distinct phases: (i) the preinvestment and implementation phase and (ii) the post-implementation operations phase. Its requirements in both phases depend on many factors, including the nature of the technology and the degree to which the phases require the transaction skills and culture of a "retail" player versus those of a "wholesale" player.

In the case of most "downstream" entities in the illustrative organizational chain where there is a group involved in the investor, operator, or user roles, there may be a requirement for a third form of intermediation, namely, social/organizational intermediation services, to assist in the creation and/or strengthening of these groups. This is especially important in certain types of rural energy investments, such as village utilities based on micro/mini hydro schemes.

The high information management intensity of renewable energy investments calls for public and public-private collaborative financing support and risk-sharing options that include help with access to information. Support mechanisms take many forms, but two deserve special mention. First are programs that provide preinvestment study support for a gamut of financing and risk-sharing terms ranging from partial to full grant, cost share, and various contingent mechanisms. If the investment goes forward, such preinvestment support becomes debt or equity-financing participation in the investment. Second are programs under which governments, recognizing that renewable energy resource information has some public good, have underwritten resource information collection and analysis and, in some instances, eased access to this information.

In some countries and for some resources such as hydro, wind, and solar, these support activities have been extensive. In some countries in Asia there has been considerable rural energy market penetration. Besides financing issues, however, lessons from this experience suggest the importance of changing markets and institutions .

Private sector players are often innovative and cost effective, both as a result of competition from other similar entities, and as a result of the pressures of competition from "conventional" options. Innovations have come in three distinct but often-interrelated areas: technical, institutional or organizational, and financing.

Examples of technical innovation include the development of the increased size, better performance, and lower unit cost of wind turbines, and the development and adaptation of low-cost microprocessor-based controls allowing cost-effective applications such as grid connection and synchronization for very small renewable energy systems and optimization of diesel renewable-energy hybrids.

Institutional or organizational innovations, including the creation and evolution of escos and other entities operating under performance- based contracts, integrate financial and technical intermediation and participate in investment ownership and/or operation. Financing innovations provide a basis for at least a subset of financial institutions to be innovative market leaders in providing and/or arranging financing for such investments.

In India, the potential for production of biogas from many industrial wastes and its high-value use as fuel for processed heat and/or electricity generation is very large. In addition, employment of this process produces significant environmental benefits. In some applications its use is sufficient to bring the resulting effluent discharge up to or close to prevailing environmental standards. The potential for cost-effective application of the technology is large due both to the presence of many relevant industrial facilities and the resultant energy and environmental benefits of the technology's application. As a consequence, a number of technology suppliers have emerged, some of them incorporating international technology and know-how under some form of technology collaboration.

Market penetration is nevertheless constrained by such factors as the capital cost and unfamiliarity with the technology. Compounding the problem is the fact that better-performing biomethanation processes often require slightly more capital and better-trained operations staff.

The result is a combined commercial and technological risk for users. In a conventionally financed project, staff training, break-in operations, and related assistance components commonly included in the turnkey provision of industrial plants can overcome the problem.

For small industrial operations, however, the capital cost barrier remains. To overcome the barriers and to aid the market penetration of its proprietary high-performance biomethanation technology, the Western India Group (WIG) has developed an innovative build-own-lease-transfer financing scheme based on the principle of shared savings. The scheme facilitates better allocation of contractual risk-commercial and performance-to the party better equipped to manage the risk.

Despite their often-attractive financial returns, many sustainable energy investments such as those for energy conservation face barriers to their market penetration. Some of the barriers are related to the investments' characteristics, others to the market realities the investments face. One inherent characteristic of energy efficiency and other sustainable investments is that they are more know-how, management, and information intensive-in planning, implementation, and operation-than conventional ones.

In the case of energy efficiency investments in commercial and industrial facilities, this characteristic is exacerbated by the scarcity and high opportunity cost of industrial managers' time and the time required to adequately investigate sustainable options. Furthermore, many investments are more capital intensive than the less efficient alternative. Thus, there is also a financing barrier that is commonly made worse by the potential buyer's financial constraints and financial institutions' reluctance to provide financing for such investments.

Energy service companies (ESCOs), either on their own or in collaboration with financial institutions, provide a "total energy efficiency package" of know-how, equipment, and, often, financing. The ESCO-customer relationship starts at the pre-investment phase and can continue through implementation and into operations. Such ESCOs are often creations of energy efficiency equipment of suppliers.

Ownership of the energy efficiency equipment either stays with the ESCO or is transferred after a period based on lease toward purchase or payments based on the sharing of the value of energy saved. Often, the contractual relationship between the customer and the ESCO has specific energy savings performance criteria to which the payments are keyed. ESCOs, especially those that provide "total packages," have been slow to emerge in the region, but there are some that lead the way.

The Chinese energy management companies are unique project developers. Unlike comparable companies elsewhere, they are in many ways an integral combination of an industrial enterprise and a bank or leasing company and require the managerial skills and business strategies inherent in both. Their financial success depends upon their ability to attract capital and bear all performance and credit or customer default risk in developing, financing, installing, and, to a greater or less degree, operating energy efficiency projects. Their profits comprise both a mark-up on the installed project cost and an appropriate return on the outstanding balance of the funds used to finance the projects, less the costs of borrowings.

In purely financial terms, the companies create and finance capital assets that are installed in, and then effectively leased to, a variety of host enterprises for varying periods. The assets generate an uncertain revenue stream from actual measurable energy savings, which is shared between the host enterprise and the energy management company in an agreed proportion and for a predetermined period of time.

In India, Bhivandi is a Mumbai industrial suburb, dominated by textile companies that operate approximately 300,000 power looms, each with a 0.88-2-hp motor. The power looms operate at a low power factor and low voltage, resulting in high distribution losses. Distribution losses in Bhivandi are about 25 percent. The use of capacitors improves the power factor and voltage profile. The maximum benefit from a capacitor is achieved if it is fixed at the load end. However, end users are reluctant to provide capacitors on their looms because they realize no direct benefits.

The cash-strapped Maharashtra State Electricity Board (MSEB) leased 300,000 0.45 KVAR LT capacitors from Asian Electronics Ltd. (AEL), requiring less initial investment than if it had purchased them. As lessor, AEL retains ownership of the capacitors and receives the benefit of their accelerated depreciation. AEL installed the capacitors at the power looms of the end-users and is responsible for maintenance. AEL provided a performance guarantee that stipulated no rental charge for faulty capacitors not replaced within seven days of being detected.

The power factor will be monitored through measurement at distribution transformers. MSEB has mandated that consumers must maintain a 0.9 power factor or else pay a penalty. Installation of the AEL capacitors improves the power factor above the mandated 0.9. The goal of installing the low-tension capacitors was to increase the power factor from less than 0.7 to greater than 0.9. Such a power factor improvement resulted in total annual loss savings of approximately 100 kWh per year, equivalent to 20 MW of installed capacity.

In addition, the leasing scheme resulted in an update of MSEB customer records. The supplier was required to keep careful records of every capacitor installation in order to receive its rental payments. The information was used by MSEB to update its customer records for unregistered power looms and horsepower.

As a result of the program, MSEB's monthly revenue has increased from Rs30 million to Rs85 million, due to normal growth; reduced interruptions as the breakage of conductors and failure of transformers are minimized; increased voltage, which has led to increased output and consumption; replacement of stopped meters discovered in the field; frequent visits to end-user sites, which helps control meter tampering; and energy audits for each distribution transformer, which helps detect and reduce energy theft.

MSEB also benefits from the payment structure: it recovers the lease payments from the consumers monthly, but is required to pay AEL only on a quarterly basis, allowing it to earn interest on the rent. MSEB created the market for the LT switched capacitors by mandating penalties for end-users with low power factors. AEL, with net sales revenue of Rs7.9 million in the fiscal year ending 31 March 1995, financed the manufacture, installation, and maintenance of the capacitors through its normal balance sheet financing. Leasing the capacitors allowed MSEB to avoid significant initial investment in capacitors when it started its reactive power management program.

End users are also able to avoid a lump-sum payment, as MSEB passes on the monthly rental charges to the end users. The program has proven profitable for AEL; increased the power factor, which resulted in loss savings; increased revenue for MSEB; and provided better-quality electric service with fewer interruptions to end users. Examples of such innovations naturally include financial institutions that have collaborated with escos and other performance contract innovators.

Many such efforts combine "promotion" with special assistance and credit programs. Examples include the considerable success in India of the multidimensional federal and state interventions to "pull" the market or to cost-share pilot, demonstration, or "pioneering" investments. At the central and state levels, regulatory changes foster "off-site" self-generation and third-party sales as well as tax depreciation and other incentives. These have led to dramatic increases in private sector renewable energy power generation.

Small-scale technologies, domestic (in-country) entrepreneurs, environmentally benign approaches, and services for and by the poor and women are normally considered, if at all, as pious intentions. It is not that the needs of women, low-income households, urban squatter settlements, and remote rural villages are unknown, unex-pressed, or impossible to fulfil. It is that they:

— are needs of the politically powerless;

— lack the scale and profit possibilities attractive to big contractors;

— are invisible to urban elites, bureaucrats, consultants, and entrepreneurs;

— are dispersed, small scale, labour intensive, and demanding of technical, financial, and organizational assistance, thus making a project a messy and unwieldy affair for technocrats and administrators; and

— involve mobilizing and empowering women, landless labourers, the socially excluded, and other groups with uncertain political effects.

The many women and children of Asia and the Pacific who depend on smoky, solid fuels (except charcoal or briquetted coal) and inefficient stoves, usually located in unventilated kitchens, are at risk. Deaths for Asia and the Pacific may total 1.5 million people annually. Rural electrification has had significant positive impact where other conditions for rural economic development were not in place, but it often bypasses the needs of women, low-income households, and even high-priority social services such as schools and health clinics.

Rural electrification is modern and technically complex. Stoves are old fashioned, culturally specific, women-oriented, and dispersed. They are also one key to transforming women's lives and improving environmental quality. Both rural electrification and improved cookstoves are needed; it is the balance between them that is out of whack. Stoves in many low-income households burn biomass fuels-crop residues, cow dung, twigs, and branches.

Burning decreases soil-building capacity and soil's nutrient content and creates air pollution. Breathing smoke from indoor air pollution caused by biomass stoves has a health effect on the cook of smoking 20 packets of cigarettes a day. Coal stoves create even more of an air pollution hazard. Efficient and smokeless stoves have significant environmental and health impacts. Furthermore, fuel saving translates into time saving for women and children as they need to spend less effort collecting biomass fuels. Children can stay in school; women can engage in more productive activities.

Substitution of biomass fuels by "modern" fuels such as lpg, kerosene, or producer and natural gas can have even greater positive impact on women, children, and soil fertility. Disability-adjusted life years is a standard measure of the burden of disease and reflects life years lost due to premature death, fraction of years of healthy life lost from illness disability, and discounted weighting of life lost at different ages.

Improved biomass stove programs are shown to have economic returns on investment of more than 200 percent, considering only fuel savings and employment generated by the programs and ignoring health benefits and reduced greenhouse gas emissions.

Fuel substitution requires assured supplies of fuel and financing for low-income households to purchase necessary appliances. The issue has many dimensions: from the Korean coal briquette stoves (yantan) that cause carbon monoxide poisoning and raise sulfur dioxide concentrations in Seoul slums to four times higher that in high-income areas, to the success of the efforts in Hyderabad, India, to encourage conversion to lpg for cooking by providing a reliable supply system to all areas of the city. The direct benefits (in terms of people's health) and indirect benefits strongly complement each other, which should encourage developing countries and donor countries to cooperate with each other in improving cookstoves and developing fuel substitutes.

In India, women and children perform 80 percent of the village's work, which means that girls are forced to leave school early and become mother's assistant. Low productivity due to inadequate investment in water harvesting, soil conservation, and tree and shrub planting are part of the problem. Inefficient stoves, low-input fodder production, and laborious food preparation are another part. From a systems perspective the potential for improvement is great.

The women and children of Syuta work so much for so little. The biomass material that they have to collect in Syuta in order to maintain their croplands' productivity and get a subsistence crop is astonishing: 1,230 tons of grass, leaves, crop residue, and manure to produce 47 tons of cereal and 19 tons of milk. The industrial revolution fathered an industrial system based on the assumption that fossil fuels would be cheap forever, that water would be free, that minerals would be either inexhaustible or substitutable, and that the disposal of "wastes" from industrial activities would be no more than a minor technical problem.

Improved Cook-stoves Program in the People's Republic of China and India

Cook-stoves programe in China was initiated in the early 1980s and covered almost 76 percent of rural households by 1996 without any direct government subsidy. Its strategies were the following:

— Work began in the areas where people showed their desire for improved cook-stoves. Site selection was made on the basis of the county's desire to become the site of a pilot program, along with other considerations such as fuel shortages and available managerial, financial, technical, and raw material resources.

— Research and development were geared toward designing and modifying stoves to match local conditions of fuel and cooking and heating needs. Local energy offices and energy companies were in charge of local activities.

— Regular, systematic, and consistent monitoring and evaluation have been an integral part of the program since its inception.

— Government contributions are small and restricted to training, administration, and promotion.

— Flow of money between the bureaucratic levels is minimal. The largest single flow that can be transmitted in a year as part of a center and county contract is only $3,000.

— Production of the critical parts of the combustion chamber is centralized to ensure reliable long-term, high-efficiency stove performance.

In India the program began in the early 1980s and covered 15 percent of rural households by 1992. The improved cookstoves were highly subsidized by the Government. The program's strategies were the following:

— Political and bureaucratic pressures led to attempts to implement the program extensively from the start. In some areas, therefore, people who were not willing to accept improved cook-stoves were provided with them anyway. In many areas, however, people ready for them did not receive them.

— Bureaucrats did not consider local people's requirements.

— Monitoring and evaluation were absent until recently.

— Government contribution is about 50 percent.

While today many agree that the assumptions may no longer be valid, the obsolete mindset holding these views continues to be dominant. Owners, managers, engineers, economists, technicians, financiers, and politicians were trained to this mindset; taught to approach problems in a reductionist manner, ignore externalities, and accept poverty and inequity; and brought up with little or no appreciation or understanding of ecology and nature.

Their mindset-the fundamental assumption of industrial revolution capitalism is the cardinal barrier to energy efficiency, clean production, and renewable energy. Culture and education dominate decision making; economics and engineering are secondary.

Study shows that energy efficiency makes technical and economic sense even in narrow accounting terms. More than 20 years of studies demonstrate the superiority of systems design, holistic thinking, and land-use planning over sector-by-sector, problem-by-problem, enterprise-by-enterprise, department-by- department approaches to production. Yet, everywhere-in a new office building, a new residential or office complex, a new pulp and paper mill, or the latest major infrastructure investment-energy is used at far below easily achievable efficiency levels. Transport is irrational and wasteful, resources are discarded as "waste," pollution levels are intensified, and the needs of the poor remain unmet. "Waste not, want not" might be one motto of a new culture, but it has too grim and stern a ring to it. It also misses the cultural excitement of new ways of improving quality of life.

Today's office building, highway, rail transit system, factory, and commercial complex are not just inefficient and wasteful, but, more to the point, ugly, unhealthy, life-shortening, depressing, and antihuman. They drain us of creativity rather than provide spaces for creativity. So often, they destroy happiness, community, cooperation, altruism, nature, and health. Holistic design and planning can start from what is fun and pleasant and create production systems that meet needs without impoverishing our neighbours or grandchildren.

Today's technology can produce buildings that use 5 percent of current energy needs, factories that use 1 percent of current water requirements and emit one tenth of 1 percent of current -residuals, vehicles that are several times more efficient and 100 times less polluting than those existing. But it will not until mindsets embrace holism and respect for others. In the meantime we are left with lists of barriers-financial, technical, and institutional.

The following are the most important ones and suggestions for lifting them: energy costs that are a small part of total costs and are typically ignored at all management levels;

— subsidized energy prices that do not reflect externalities;

— high initial costs, mismatch of incidence of investment costs and energy savings, high user discount rates, and lack of access to credit for energy efficiency investments;

— priority on investing available capital in areas other than energy efficiency;

— few tax benefits, rebates, and subsidies for use of energy-efficient technologies;

— higher perceived risks of more efficient but less familiar technology;

— lack of trained personnel or technical and management expertise;

— lack of information about available technologies;

— abandonment of the highly effective energy quota system in centrally planned economies;

— poor quality of electricity supply that militates against less robust but more efficient lamps and motors that are damaged by wide voltage fluctuations;

— limited availability of energy-efficient equipment in the market and lack of comprehensive and enforced standards;

— widespread corruption in energy supply organizations that distorts incentive structures for energy saving;

— high transaction costs; and

— regulatory restrictions that make it impossible for enterprises saving electricity to sell savings to neighbouring enterprises.

The semiconductor industry is a major consumer of energy, water, and other materials. The processing power of microchips doubles and prices halve about every 18 months, but to achieve this incredible performance, engineers have neglected the lesson of chip design and built plants that are highly inefficient.

6

Sustainable Future

The lack of assets, opportunity, and effective voice for large segments of the population blocks the emergence of general welfare-enhancing policies, impedes growth, and undermines the potential for positive change. A more sustainable development path is more socially inclusive; it enables societies to transform and solve collective action problems. The challenge, now and in the future, is to develop the courage and commitment to manage the processes that underpin human life and well-being and to bring about a transformation that improves the quality of the environment, strengthens our social fabric, and enhances the quality of people's lives.

In many developing countries, productivity is low, growth is stagnant, and unemployment is high. The number of people living on less than $1 a day is dropping but it is still a challenge, and more people are living on fragile lands. Income inequality is rising. Average income in the wealthiest 20 countries is 37 times that in the poorest 20 countries-twice the ratio in 1970. Many of the poorest countries are wracked by civil conflict, with animosities deep and prolonged. Stress on the environment is increasing.

Science and technology are will not be sufficient without complementary changes in institutions. The world community, in confronting some of the challenges, is grappling with new strategies and goals. Development agencies are shifting to more participatory and holistic approaches with a medium-term perspective, through the Comprehensive Development Framework (CDF) and focused poverty reduction strategies, backed by actions on the ground through partnerships, broader inclusion in the preparation of assistance strategies, and some shifts in lending and grant aid. The private sector is more committed to sustainable development, with greater use of triple-bottom-line accounting by firms and greater use of environmental and social criteria by investors.

Governments and civil society are supporting the Convention on Biodiversity, the Convention to Combat Desertification, the Convention on Climate Change, and the Millennium Development Goals. And civil society is demanding more public and private accountability. Although encouraging, these advances are small relative to the many challenges of sustainable development. Sustainability requires thinking long term, but acting now-it also requires coordination.

Technologies and preferences will change, and resource allocation and distribution issues cannot be neatly separated. Managing risk is important. Not all risks are insurable. Not all irreversible changes are bad, but some are. For the assets most at risk-the natural and social-markets cannot provide the basic coordinating functions of sensing problems, balancing interests, and executing policies and solutions. These types of assets have impacts, good and bad, that extend beyond individual transactions and thus require coordination to promote good spillovers and minimize the bad. The rising income can facilitate but not guarantee better environmental and social outcomes by permitting countries to simply "grow out of " pollution or civil conflict.

Problems with impacts and risks that are diffuse and long term (such as climate change and biodiversity loss) are less readily perceived and appreciated than those immediately felt and measured (i.e., some forms of local air and water pollution and deforestation). Assets that are public goods or common property goods-such as clean air, forests, fisheries, and water-are a challenge to manage sustainably.

The reason is that private property rights are difficult to assign or enforce, or if improperly designed they fragment and undermine the underlying joint functions of an asset-say, an interconnected ecosystem. In addition, the concerns of the poor and powerless are less likely to attract the attention of society's many institutions than those of strong and vested interest groups. That is why greater equality in access to assets and voice makes a difference in the kinds of environmental and social concerns that society addresses.

When the poor lack voice and a stake in society, social assets (such as trust) and environmental assets (on which the poor depend) are eroded, stability is undermined, and the ability to solve economic, social, and environmental problems dissipates. At the very least, inequality and the lack of hope for the poorest countries will inhibit attempts at solving global problems-not just the current preoccupations with cross-border spillovers of conflict and terrorism, but also the currently unimagined problems that will require global cooperation 20 or 50 years hence. That is why ending global poverty is much more than a moral imperative-it is the cornerstone of a sustainable world.

In the future, we will also witness completion of the urban transition in most countries. The demographic transition will permit greater savings if the working age population has jobs and investment opportunities. The urban transition will facilitate income generation, but over the next 50 years it will also require massive investments. Many of these investments are likely to be long-lived-so getting them right by incorporating environmental and social concerns in their design now is critical. Even the next 15 years could bring a record period of economic growth in developing countries. Driven by growth in China and India, income in the low- and middle-income countries will almost double-accounting for more than a third of the 60 percent increase in world output.

This period offers the opportunity to lay the foundation for inclusive growth-which will require confronting barriers to change. Institutions that can manage the social and economic transitions, by partially compensating losing interests, are much easier to create in rapidly growing economies than in economies where inclusion requires a battle over stagnant shares.

For many countries whether the opportunities generated by new growth are inclusive, or whether they lock in vested interests and exclusive institutions, will depend on decisions taken in the next few years. Sustainable development requires action across many sectors and disciplines, including water, energy, health, agriculture, biodiversity, and others. In a companion document, the World Bank sets forth some recommendations and action plans to advance sectoral goals.

Achieving all the broad sectoral goals will involve problem diagnosis, decisions with distributional consequences, and coordinated and sustained commitments to action. These functional capabilities require general improvement in coordinating institutions within and across countries. Making progress on the sectoral issues, for example, requires first a better understanding of local conditions and a better ability to diagnose local problems.

Domestic and international institutions will be required to fund and implement R&D where local capacities are deficient. Health and water services often deteriorate after costly initial investments for lack of routine funding and maintenance. This speaks to the challenge of organizing beneficiaries for sustained commitments.

The village initiative in Morocco is being replicated in a dozen other villages. The forest concession program in Cameroon has created a constituency for expanding it to other sectors. There is much creativity under way with initiatives emerging from the public sector, the private sector, and civil society. An enabling environment is needed to encourage such initiatives, to facilitate partnerships, and to help mobilize the resources needed to scale up promising activities within countries, and across countries.

In the early stages of the epidemic, HIV/AIDS received little attention since no one knew how contagious it was or the trajectory of future costs. By now, with the evidence available, all countries should have programs to identify the problem early and to stay ahead of the epidemic. The same is true for environmental is-sues-the status of biodiversity, air pollution, lead in gasoline, or the drawdown of acquifers in arid regions.

There is a need for significant investment in information and indicators at global and local levels, where this information would find users and audiences. The initiative must go well beyond current attempts to devise indicators of sustainable development, including summary indexes that try to capture sustainable development in a single headline number.

If society is to sustain development, it needs good indicators, but it also needs individuals and groups who demand and use detailed, and quality information to solve problems, and others who produce information to meet that demand. Environmental and social problems are easy to misdiagnose when there is a lack of reliable, current, and geographically disaggregated information. For many important aspects of the environment, global conditions or trends are worrisome, but society lacks the detailed data to monitor, diagnose, and manage the problems at local, national, and global levels.

For most countries, reliable, up-to-date, spatially desegregated information is lacking on poverty and many other social concerns (health, education, crime). These data gaps inhibit understanding of-and consensus on-the impacts of policy reforms, national and international, on poverty in the developing world. They also impede the formulation and execution of strategies to combat desertification, other forms of ecosystem degradation, and biodiversity loss. And they muddy the discussion of the nature and impacts of global inequality. Fortunately, rapid changes in communication and information technology make it more feasible to gather this information through a combination of surveys, reports from ground observers, and information from satellites and other sensors.

As information costs continue to plummet, the scope for expanding the collection of this kind of information is immense. Innovative information systems can track and deter industrial polluters and illegal deforestation and forest fires. Although information provision can sometimes be a catalyst for change, the most effective systems are those that create constituencies for information. The constituencies may be public agencies that use the information to plan and assess strategies for pollution control, crime prevention, or public health delivery. Or they may be civil society organizations that use information to mobilize dispersed interests. It is demand by users that stimulates and maintains the production of relevant, reliable information.

To equitably and efficiently balance interests within a society, two elements are necessary: first, getting everyone fairly represented at the bargaining table; second, facilitating negotiation once everyone is there. For many problems at the level of the community or nation-especially those relating to environmental and social issues-the responsibility lies with the government.

But governments vary in their capacity to undertake these tasks. An expanding set of institutional tools can assist or complement governments in balancing interests, and they can assist citizens in ensuring that their governments are fair and responsive in doing so. These tools become even more important at the international level, where they must substitute for government. It is fundamentally difficult to balance interests in heterogeneous societies. But taken together, in a context of increasingly democratic institutions and decreasing costs of information and communication, these tools give some hope.

Providing this kind of information helps level the playing field for negotiation, since less powerful interests will typically have poorer access to information. Performance reporting can also help governments, companies, and other organizations understand the effectiveness and impacts of their own actions, for instance National reporting-transparency in fiscal affairs, reliability of legal institutions, and adequacy of environmental impact reviews-can all provide the information and incentives necessary to improve governance and the balancing of interests.

Examples include the recent International Monetary Fund Codes of Good Practice on transparency in fiscal, monetary, and financial policies, and WTO requirements on subsidy reporting. Another example is the Aarhus Convention-a voluntary commitment to environmental transparency. National reporting requirements under the WTO, multinational environmental agreements, and other international treaties can help nations track and manage their own compliance processes and help build confidence among parties to forge agreements.

Independent audits of public programs-especially programs related to the management of public assets, such as land, water, forests, minerals, and fisheries and the regulation of pollution-can deter corruption and promote better management of environmental resources. Regular audits of national environmental assets can be useful in detecting and diagnosing problems. Public ratings of governments' transparency and corruption can affect investment and provide a check on government capture by vested interests.

Governments and civil society can seek ways to facilitate negotiation between affected parties. For instance, Colombia's regional environmental authorities have set up structured negotiations between water polluters and water users to help determine tolerable levels of pollution. Participatory budgeting has transformed the budgetary process in more than 80 Brazilian cities.

At the international level, the Convention on Long-Range Transboundary Air Pollution forums on air pollution and the Global Environment Facility-sponsored transboundary diagnostic assessments for international waters provide structured means of assessing options. These forums become more effective when they build up social capital through dense networks of trust and information linking technical experts, government officials and legislators, civil society, and special interests. Compensation and incentives. Even win-win outcomes usually have a loser-a party whose losses are outbalanced by social gains.

Basic principles for balancing interests therefore include minimizing the losses, compensating losers, or providing incentives that reconcile private and social objectives. One way to do this in the environmental sphere is through market-based permits. Where actors have different costs of complying with social objectives, trading rights or obligations can substantially reduce compliance costs for individuals and for society as a whole. Assignment of valuable permit rights is also a means of compensating potential losers and ensuring participation in a reformed system.

Certification systems are another means of aligning private and social incentives. Certification of firms helps investors identify companies with better environmental and social performance, thus promoting incentives for more socially responsible behaviour. Certification of financial institutions can promote standards for assessing the environmental and social impacts of investment projects. Certification of products (wood, coffee, fish, beach resorts, garments) can allow consumers and investors to reward firms that employ environmentally and socially sustainable production processes. Certification of diamonds and other lootable commodities can be part of a strategy to avert civil conflict over point-source wealth.

Even governments may find it advantageous to participate in voluntary certification schemes. For instance, the Alaska state government applied for, and received, certification by the Marine Stewardship Council for management of its salmon fishery. The success of these efforts depends on how they influence producer, consumer, and investor behaviour. And that depends on the costs of certification, the proportion of the industry that is sensitive to certification, and the legitimacy, integrity, and reliability of the certification criteria and process. To become more widespread and effective, certification systems require coordinated international attention on several grounds. First, there are important links to the world trade system and a need for coordination with WTO activities-to ensure that standards are not used to disguise protectionism. Second, careful attention is needed to monitor the possibility that complying with certification procedures may be too burdensome for small firms. Third, and probably most important, there is a need to ensure the integrity of the evolving certification systems.

For these systems to work on a large scale, they will probably require a massive expansion of "markets for honesty." These are the networks of private auditors and certifiers, and their accreditation agencies, that provide ISO certification, shipping certification, and audits of corporate earnings. Implementing and executing policies and programs requires appropriate institutional capacity.

Mention of capacity-building evokes respect and approval-but not always excitement or enthusiasm. It has usually been seen as an add-on to projects and programs with other, more important purposes. It is often identified with failed technical assistance projects that relied heavily on the ephemeral input of foreign consultants. The same is true of monitoring and evaluation, another project add-on that often fails to be executed with rigor or to provide much insight.

Promoting capacity-building and problem-solving in the developing world. Yet, despite technical cooperation grants of $201.3 billion over 1990-2001, 6 progress in institution-building in developing countries has been disappointing. Most of those funds, to be sure, have been devoted to project preparation. Such investments may have been effective in achieving project goals through reliance on expatriate inputs, but at the cost of forgone opportunities to bolster local capacity. There are failures, too, in the demand for institution-building. Often, lack of ownership, lack of government interest in bolstering capacity, and pressures for rent-seeking have kept competent institutions from taking root.

Several new approaches could help place learning and institutional development at the core of sustainable development efforts-that is, projects and programs designed around institutional development, rather than vice versa. These approaches recognize that capacity-building is not accomplished in a few years and thus is not well suited to lending for individual projects or adjustment programs.

These approaches also recognize that capacity has a strong social capital component, involving not just people but enduring networks that link policymakers, experts, civil society, and the private sector. And these approaches generate virtuous circles when information and indicators are made available to groups that demand and act on that information. The new approaches use a variety of new types of learning organizations to stimulate both the demand for and supply of institutional skills; they include:

Scientific research organizations, based in the developing world, that combine local and global expertise and help to nurture "policy entrepreneurs" at home in both scientific and policy worlds. Learning networks for sharing knowledge, such as the Union of Capital Cities of Ibero-America-real-time, demand-driven learning networks run by national and international associations of mayors, local government officials, and city practitioners share experiences among their members and other cities through the Internet and workshops.

These groups that independently monitor and report on government and corporate expenditures and activities have proliferated recently. They can help improve government and corporate accountability, create incentives for monitoring and evaluation within government and corporations, and mobilize pubic demand for institutional improvements.

Donors and multilateral development banks can support the evolution of this intellectual ecosystem of organizations that learn-and apply that learning to improving policies and projects. Donors can support these organizations, through direct funding and twinning arrangements on a large scale. These are long-term efforts that bear fruit over a decade or two as the institutions train people, enhance the prestige of necessary but neglected professions, such as policy analysis, and build dense networks of trust and knowledge.

Funding must be committed over periods much longer than traditional projects, and funders must accept that the impacts of these investments, though potentially enormous, will be deferred and difficult to quantify. Donors can also design projects with learning as a central output, providing hands-on monitoring, evaluation, and implementation experience to learning organizations.

At the transnational and global level, assessment institutions such as the IPCC have shown their value in forging consensus on the problems and the options for addressing them. More institutions like the IPCC are needed to address the new global problems that continue to emerge all the time. For instance, questions at the intersection of trade and environment are sure to proliferate, as the discussion of product certification illustrates.

Trade policy is an area where more systematic analysis of options might help in forging agreements. Deepening scientific knowledge exposes overlooked transnational environmental processes. For instance, there is increasing attention to the global nitrogen cycle and its effect on marine ecosystems. Technical change, too, brings new problems and risks as well as opportunities.

Balancing the environmental risks and benefits of genetically modified organisms is a clear example. And social changes require ongoing and forward-looking attention. For instance, intensifying pressures for international migration have far-reaching ramifications. In all of these areas, there is a strong argument for concerted international attention-and for achieving some consensus on the relevant issues. Ensuring greater inclusion Increased voice and major increases in substantive democratization.

Inclusiveness can be expanded through significant changes in governance that increase representation and accountability, such as empowering local government through well-designed decentralization reforms; electing rather than appointing mayors; replacing military with elected regimes, or empowering groups excluded from decisionmaking-women, indigenous people, and other disadvantaged groups, who may be in the majority.

Better distribution of access to assets. Dynamic growth and development processes create more assets and new types of assets. It will be much easier to increase inclusiveness by ensuring that the poor and disenfranchised, as well as the middle class, have greater access to these newly created assets. By increasing access to education, which build human capital; by expanding market-based rural land reforms to increase smallholders' access to agricultural land and complementary assets (water, roads, and know-how); by expanding the provision of secure tenure (protection from arbitrary eviction) in urban slums or other informal urban settlements; by increasing access to knowledge (the new asset frontier).

To overcome the barriers to solving collective problems more rapidly and systematically requires mutual commitments by developing and developed countries to a bold global vision and accord. This vision requires a massive and steady effort to eliminate poverty and to protect and manage a broader portfolio of assets that will ensure the well-being of future generations.

The European Union shows how the prospect of mutually beneficial integration can induce poorer countries to adopt higher standards of environmental and economic management while the richer ones provide resources and help to boost capacity. The Dust Bowl in the United States shows how small, individual states could not solve their problems without the migration opportunities offered by other states, or the channelling of knowledge and financial resources from other states that helped to restore economic health over many decades. The European Recovery Program (the Marshall Plan) after World War II showed how mobilizing resources on a grand scale can build economies and transform enmity into partnership. The architects of the Marshall Plan accepted the challenge of tackling "hunger, poverty, desperation, and chaos" by rebuilding a continent in the interest of political stability, social development, and a healthy world economy.

The Treaty of Versailles ending the First World War in 1919 had imposed unilateral conditions and enforced severe reparations on the vanquished, paving the way for political extremism. The designers of the Marshall Plan avoided these mistakes and paved the way for peace. The Treaty of Versailles courted conflict. The Marshall Plan broke a vicious cycle of poverty and regret; it supported economic reconstruction and social order; and it injected money and ideas to rebuild Europe and herald more than 50 years of unprecedented peace, prosperity, and partnership. Balancing interests and forging credible commitments for the long haul is difficult at the national level but even more so at the global level.

So increased global coordination is necessary to expand the capacity and opportunities of the weaker segments of the global community. A self-enforcing global accord may be required to get the commitment to finance such a scaled-up effort, to build capacity to use the funds wisely, and to take on difficult reforms in developing as well as industrial countries. There is growing recognition of the need for mutual commitments and for accelerated improvements in key development indicators.

Support for the Millennium Development Goals, which propose to cut the proportion of people in extreme poverty by half by 2015 is now widespread. The goals set ambitious quantitative targets for reducing hunger, increasing primary schooling, improving health, promoting gender equity, and ensuring environmental sustain-ability. One calculation puts the cost of meeting just the nonenvironmental targets at $40 to $70 billion a year above the current $50 billion in development assistance.

The International Conference on Financing for Development in Monterrey also confirmed the need for more aid, trade, and debt forgiveness by industrial countries, in tandem with domestic reforms in developing countries to increase domestic resource mobilization and facilitate foreign direct investment.

The proposal for the New Partnership for Africa's Development (NEPAD) is also based on an arrangement in which developing countries take responsibility for improved governance, and industrial countries help through more aid, debt forgiveness, and market access. The Council of the European Union's proposed Global Deal has many of the same elements.

Many of these long-horizon initiatives will yield benefits in the medium term in support of the Millennium Development Goals. But because many future problems cannot be foreseen, it is important that institutional foundations be strong and that a process and a framework be developed that are robust in picking up new signals, balancing a broader range of interests, and maintaining commitments to the global vision.

Most of the physical capital required for the economy 50 years hence has not yet been created. This provides an opportunity to incorporate inclusiveness and sustainability criteria in new investments now. The potential is there to shift development paths, provided institutions that adapt and implement better policies can be put in place.

Investments in global poverty reduction and in greater inclusiveness will help ensure the representation of all interests in the design of the new and improved institutions. There is no understating the difficulty of these challenges.

Social inertia is great, and institutional change can take decades. Overcoming the inertia to tackle these difficult problems-the fears and risks of unilateral action-requires coordination. Providing the necessary funding. Capital markets can cover much of the funding required to shift to a more sustainable path if appropriate policies are in place. For example, they can cover investment in new and replacement capital to improve energy efficiency and meet the demands of an urban population that will double. But expanded domestic resources and development assistance will be needed to cover the part of these costs that involve the provision of local, national, and global public goods. Institution building is one of these public goods.

The estimated funding requirements for the Millennium Development Goals, would not be enough to support a broader and deeper agenda of institution-building. A development strategy that emphasizes inclusiveness, shared growth, and better governance places large demands on leaders in developing countries. They must commit to better economic, social, and environmental management-and thus to better governance.

Schooling, health care, and provision of environmental assets that protect health, market-based rural land reform, and regularization of urban tenure all promote asset generation for the poor. An open and verifiable flow of information is important to tighten accountability in government and the private sector through such steps as opening procedures for bidding, strengthening meritocracy in the civil service, and making sure that public and corporate budgeting and resource management are governed by law, open to the public, and under proper oversight institutions.

The success of the reforms would be long term; they would secure opportunities and voice for families and their children so they can save, invest, and engage in their communities. But good policies, to be sustained, will require committed support and the legacy of reversals and stalemates is stunning. Research on aid and policy shows that aid is not worth much without good policies and institutions; in fact, it can even be harmful. Research also shows that good policies and institutions are essential to growth and development.

Funds and assistance will not be available without the conviction that there is capacity to use them effectively. But capacity-building requires patient investment to remove critical barriers because its pay-offs are large but take time to be realized. Fundamental reforms require better institutions, which evolve slowly. These requirements are interlinked. So developing country leaders need to know that they will have long-term, reliable support from the larger development community. Without such support-and the quid *pro quo* on reform to support it-many developing countries will remain in cycles of promise and disappointment. With support, these cycles can be broken.

Enabling developing countries to develop more rapidly through increased aid, trade, migration, and access to knowledge and technology will place big demands on leaders and voters in developed countries. Developed countries should strengthen the ability of developing countries to pursue sustainable development by providing development assistance that supports public goods and attracts private investment.

Developed country trade barriers impede exports from developing countries and undermine the livelihoods of the poor. Unrestricted access to developed country markets in textiles and clothing could yield $9 billion a year, and access to agricultural markets $11.6 billion a year. Improve developing country access to technology and knowledge. Implement incentives to promote the transfer and dissemination of technologies to developing countries-including those for climate mitigation, disease prevention, and agricultural development. Together the developed and developing countries must address the most urgent problems facing humanity.

Their joint responsibility is to establish a global partnership to set the rules for making rules and the modalities of burden sharing. As the world becomes more interconnected-environmentally, economically, socially-new institutions and rules must be agreed on and implemented. These will include rules for international trade; rules to avert conflict; rules on migration; rules governing the use of the biosphere; and rules affecting property rights in ideas, technological processes, and genetic information. The consequences will be enduring.

Governments of developing, transition, and developed nations; provinces, cities, and local communities; civil society organizations; private firms; individuals. All governments can improve the accountability of public agencies and the provision of information about social and environmental conditions-to improve the ability of the general public and civic groups to identify problems, balance interests fairly, and come up with solutions. Civil society organizations can help to aggregate the voices of dispersed interests and provide independent verification of public, private, and non-governmental performance.

The private sector can advance economic, social and environmental objectives by helping to construct a framework that provides appropriate incentives for firms to be accountable in all three dimensions. If the global community sees merit in such an accord, the accord's elements will need more careful work over the next few years to develop an implementable program that can adjust to contingencies without undermining the promise of the accord.

One interpretation of overconsumption is that it refers to the environmental externalities associated with consumption at higher levels of per capita income. For example, carbon dioxide emissions, and their contribution to climate change, are highly correlated with consumption of electricity, home heating, transport services, and energy-intensive manufactured goods-all of which tend to increase strongly with income. In these cases, the over prefix is justified, since the externalities are by definition inefficient and usually inequitable. But the overall level of consumption is not the source of the problem. It is the combination of the specific consumption mix and the production processes that generates the externality.

Another interpretation of overconsumption, much more difficult to document, has to do with social externalities. People judge the adequacy of their consumption-clothing, automobiles, housing-in part against norms set by others. If this is true, consumption takes on some of the aspects of an arms race. Mutual restraint is needed to shift resources from competitive individual consumption to consumption of public goods. But these externalities need to be much better understood before there can be any agreement on the actions to address them. Despite great promise for improving the agriculture of the poor, biotechnology in general and transgenics research in particular have barely begun to address the problems of the poor.

Comfort with the new technology is determined in large measure by societies' comfort with their scientific and food safety institutions and their feelings about emerging concentrations of economic power in multinational "life-sciences" corporations. Solutions to these complex issues are all playing out against a backdrop of globalization-related uncertainty, which has left many people unsettled over their capacity to control their lives and their environment. It is the rural poor in developing countries who most need access to these new agricultural technologies.

Intellectual property rights (IPRs) is the next generation of assets that can increase or decrease inclusiveness with consequences for the evolution of quality institutions. IPRs represent a compromise between the interests of users, owners, and creators. It often costs very little to duplicate a seed, a computer program, a song, a drug, a blueprint for a printed circuit or a meteorological database. Once these products have been created, their widespread dissemination would bring great consumer benefits. Intellectual property rights such as patents and copyrights, balance these static and dynamic aspects of efficiency.

The Trade-Related Aspects of Intellectual Property Rights (TRIPS) agreement under the WTO represents a global strengthening of the rights of producers in terms of users. Its immediate effect will be to increase royalty payments to intellectual property rights holders, who are overwhelmingly in the developed world.

In principle, TRIPS provides a wide latitude for a developing country to fine-tune an intellectual property rights system appropriate to its needs. In practice, developing countries' ability to manoeuvre may be more limited, and the potential for unequal outcomes is worrisome. While the outlines of the global intellectual property rights regime are clear, many crucial details may not yet be established.

Global inequality, combined with global demographic trends, will create more pressure for migration. An extensive literature shows that wage differentials drive migration. Fertility rates are highest in places that have the least capacity for absorbing labour. Meanwhile, aging in the developed world will drastically reduce the size of the labour force and increase the demand for low-skilled labour for tasks resistant to automation, such as care of the elderly.

There are many reasons to support both more long-term as well as circular migration (the latter can help speed up learning in lagging regions), but migration remains a politically sensitive issue in receiving countries. There are issues related to assimilation of newcomers-as well as exaggerated fears and misconceptions. The decision of one potential receiving country to restrict immigration has implications for other receiving countries, and for the sending countries. Stresses associated with immigration may be related more to the rate of change than to the level of change.

Sustainable Agriculture Development

Sustainable development requires altering social relations and recognizing the intensely political nature of environmental change and poverty. It transcends economics because economics has failed to recognize the fundamental processes of nature, the commodification of nature, social struggles to claim rights over nature, externalities arising from the exploitation of nature and labour, and the depreciation of nature by such exploitation. Sustainable agriculture does not require merely curbing the more egregious excesses of high-input industrial agriculture controlled by marketing and input-supply cartels. It also requires scrapping the system itself and moving in a revolutionary direction toward a holistic or systems approach to agriculture. Systems thinking revolves around social relations and nature. It seeks to affirm the power of a mobilized society and to establish a new, healthier balance among State, market, and society.

Sustainable agriculture implies new ways of approaching rural development. It is not so much a criticism of past approaches as a recognition of the shortcomings of narrowly focused technocratic solutions. That rural regeneration toward sustainable agricultural and rural development has at least 12 aspects. The 12 aspects are the following:

1. self-help mobilization and the creation of many loci of action in rural areas;
2. improved governance and support to catalytic community organizations;
3. constant affirmation of political and human rights, which involves struggle and conflict;
4. constant vigilance and government intervention to correct market imperfections;
5. political mobilization of the disempowered to target public services more precisely;
6. provision of access to land to reduce poverty and promote equity;
7. making common property resource management a focus of policy dialogue and not merely an academic sideline;
8. recognition of water management and allocation and, especially, irrigation management as key to sustainable use of water resources;

9. targeting financial services in less-favoured agricultural areas;
10. forming cooperatives, which is essential to reinvigorating smallholder agriculture and creating off-farm employment;
11. State support for appropriate R&D and extension services aimed at less-favoured agricultural lands, smallholders, women, and sustainable agricultural practices; and
12. employment guarantee schemes to reduce vulnerability.

The people's organizations must share interests but not be predicated on the assumption of a homogeneous "community" of shared interests. They are bound to clash with local elites over access to and use of local resources. They must therefore pay careful attention to dispute resolution that recognizes asymmetries of power and strives for equity and justice in outcomes. Water-users' groups, forest-users' groups, production and service cooperatives, micro-credit borrowing groups, landless water sellers, and equipment and work-sharing groups, among others, are already active in many countries' attempts to mobilize farmers for livelihood improvement.

They have encountered many difficulties, perhaps the most serious of which has been their takeover by the elite and powerful who use them to obtain subsidized inputs, retain control over resources such as irrigation water or timber, maintain coercive social relations, or head off organizing by the poor. Fortunately, many groups have managed to remain cohesive, operate efficiently and effectively, and build the power of the poor, women, and other socially excluded groups.

Sometimes organizations benefit the communities as a whole, especially those that are more homogeneous, less faction ridden, less stratified, under less pressure from outside market or State forces, and have rural classes with common interests.

The catalytic organizations' goal is to create many loci of action throughout rural areas. They will sometimes oppose and sometimes support State interventions, but they will always seek to mobilize local creative energies expressed as ideas, traditional knowledge, labour, savings, locally available resources, and aesthetic sensibilities. Their efforts are motivated by the belief that it is necessary to wean people from the idea that the State can and will solve problems of poverty, environmental destruction, social dislocation, and social exclusion.

Catalytic institutions such as people's organizations, user groups, NGOs, and co-operatives flourish in the fertile and sustaining soil of democracy, which respects human and political rights, has high standards of government accountability and transparency, and listens to and responds to the pleas of the powerless. Decentralization does not mean divesting higher levels of government of responsibilities that are rightfully theirs, but rather devolving responsibilities and powers to the appropriate level for efficiency and effectiveness.

Environmental regulations, for example, will remain the responsibility of nonlocal authorities, but monitoring and sometimes enforcement can be delegated to accountable local authorities. Reducing Poverty-Findings and Implications reveals that decentralization of authorities and resources to the local level, which weakened governance, has increased poverty. Therefore, such decentralization must be accompanied or, preferably, preceded by establishment of good governance and capacity building.

Agricultural knowledge, for example, has too often been assumed to be the exclusive realm of the outside expert. More and more, however, it is recognized that sustainable agriculture requires a process of mutual learning between the farmer and the expert and that both can contribute to finding new and better ways to attain sustainability.

Barriers to equal access to land, credit, extension, education, and health care need to be removed. At a more profound level, however, is the recognition that women's exclusion from resource access stems from inherent structural and institutional discrimination. Change at this level requires mobilization, organization, collective action, and increased political representation of women or other socially excluded groups. Legal rights enable such positive change to happen.

Price distortions are caused by implicit or explicit subsidies on fertilizer, fuel, electricity (including that used for irrigation pumping), pesticide, and credit. Carefully targeted subsidy elements-such as grants to institutional development and group organizing of rural financial institutions-can be non-distorting and non-crowding to competition if correctly designed. Incorporate in prices additions that reflect the external costs of use of the commodity. For example, electricity and fuel prices should reflect not only the marginal cost of production and delivery but also the externalities imposed on society and the environment by production and transport.

The needs of environmental sustainability, women, the poor, and the socially excluded through agricultural extension, agricultural research and development (R&D), primary-school education, adult literacy training, other nonformal education, user-group formation, local government strengthening, infrastructure investments, and rural development projects in general.

In 1990s, Philippines, the majority of farmers of the lowlands of Cotabato province in Mindanao were caught in a debt trap. After two decades of embracing the high-input chemical-based green revolution technology package of chemical fertilizers, chemical pesticides, hybrid seeds, and monocropping system, environmental degradation finally caught up with the farmers. Years and years of wanton disregard for the "ecological capital" of production in pursuit of high yields now demand payment.

Land and other properties have been lost to agricultural traders and chemical input dealers, to financiers and moneylenders. Unable to pay for their production input credits, farmers were forced to mortgage their lands and household appliances. To combat the cycle of debt and poison, many NGOs and people's organizations in the Philippines have turned to one form or another of sustainable agriculture, emphasizing diversification and little or no use of chemical inputs. One such NGO is Don Bosco, which now has over 1,400 farming families practising chemical-free agriculture in Mindanao.

Formerly deep in debt, with an ever-increasing need for chemical fertilizers and pesticides, and their land mortgaged to pay for debts incurred to pay for their production inputs, they were barely able to feed their family. Now, they can send their children to college, have redeemed their mortgaged land, paid their debts, and have a secure food base of rice, vegetables, and fish. With their savings they have also started an off-farm economic enterprise-a sari-sari store. There are many stories like theirs as biodynamic agriculture spreads in their municipality and beyond. Everywhere in the region, austerity is meant only for the less powerful. Cuts in public services must therefore be restored, although mere restoration and increase in public service budgets are by themselves insufficient to increase services relevant to sustainable agriculture.

Some countries, mostly those in South Asia, must be committed to providing primary education for all, which requires improvement in quality and not only financial increases. Better quality requires more government accountability, guaranteed rights for women, a citizenry willing and able to challenge poor school performance, and access by all to education.

Countries such as India, Philippines, and Pakistan, which have refused to acknowledge the importance of landholding patterns on both equity and productivity, have failed to reduce rural poverty. All countries, however, need to continuously re-examine their land distribution patterns, land markets, tenancy relations, gender rights and restrictions to land use and ownership, guarantee of indigenous systems of land stewardship, and the allocation of land among State, community, and private ownership, to assess how allocation can be improved. They require a local approach-one sophisticated enough to distinguish among land management and ownership rights, tree and other product rights, and water rights. While democratic community user groups should exercise most of the control, they may co-manage resources with the State.

Water is essential to promote health and agriculture. Water management and allocation have strong equity dimensions. Priority must therefore be given to radically improving the efficiency of water use. Ninth Aspect Providing financial services to residents on marginal lands, the poor, women, and fishers. Such services are essential to promoting environmentally sound resource conservation and management and to improving the livelihoods and capabilities of powerless groups. People need credit, secure savings, and insurance- crop insurance, disaster insurance, and burial and marriage funds-which can greatly help households face risk and broaden their agricultural production possibilities.

Rural areas are characterized by imperfect markets for inputs and outputs. The resultant fixed transaction costs are negatively biased against small farmers. If they must purchase and sell alone, they are forced to buy dear and sell cheap. Some technologies may be out of their reach altogether.

Agricultural production and service cooperatives are a primary means to overcome these market biases and empower small farmers or other groups of rural disadvantaged. Cooperatives have faced many difficulties in Asia over the years, but the successes such as the dairy and oilseed cooperatives of Gujarat, India, show the importance of cooperatives in a strategy that fosters smallholder agriculture. Cooperatives offer the possibility of capturing scale economies in production or processing of agricultural commodities. Capital-intensive technologies become available to small farmers.

Marketing power and competitiveness increase for cooperatives that can buy in bulk, negotiate with powerful outside buyers and sellers, improve access to credit, and save substantially on transaction costs. Cooperatives offer the possibility of improved access to technologies, markets, and knowledge about marketing and production opportunities. Cooperatives reduce the risk for individuals and thus open up many opportunities currently the exclusive domain of rich farmers. Giving up on cooperatives dramatically tilts the playing field in favour of the already advantaged.

In the future, smallholder agriculture depends upon co-operative formation and that cooperatives are one way to foster competitive, efficient markets in many agricultural services. Otherwise, the pressures that lead to oligopoly and monopoly and small farmers' limited access to information will lead inevitably to concentrated market power. The many different types of cooperatives should be investigated for their utility, not only in production but also in input supply, marketing, and information dissemination.

Research and development are the key to sustainable agriculture. If agriculture is to move toward holistic resource management, sophisticated mimicry of natural processes, reduction in expensive external inputs based on cheap energy, and long-term maintenance of soil and water qualities, then agricultural R&D must focus on these needs and on the knowledge and requirements of the small farmer. There have been successes and it is important to build on them. Agricultural R&D needs to abandon the silver-bullet approach that seeks a one-size-fits-all solution to agricultural production. This approach yielded dramatic results on favoured land and for favoured farmers in the green revolution.

Sustainable agriculture, however, demands an approach tailored to specific conditions, less-favoured areas, and less-powerful people. The nuances of how technology dovetails with social relations, agro-ecological conditions, and cultural norms need to be a part of R&D based on local knowledge and conditions. Agricultural market institutions intersect with social relations to segment markets in numerous ways, including by excluding women. Women, for example, are often restricted to localized, small-scale, lower-risk, and less-remunerative forms of trade than men.

In the year between 1995-1997, the first set of 50 people's biodiversity registers and covering 50 village clusters, distributed over seven states in India. They were prepared by college students and teachers, NGO workers, or forestry officials after extensive discussions and field visits with local people. They involve mapping the landscape and delineating the different user groups in terms of their relationship to the natural resources. The processes are followed by systematic group as well as individual interviews and field visits involving representatives of different user groups and individuals particularly knowledgeable about living resources.

Particular attention is paid to the reconstruction of the recent ecological history focusing on changes in the landscape as well as in populations of plant and animal species. Of particular relevance are patterns of access to and harvests of wild as well as cultivated living resources, and the appropriation of the benefits of their utilization by different segments of local society and by outsiders. Continuance, erosion, or emergence of practices of sustainable use and conservation of biodiversity, local as well as State sponsored, are carefully assessed and related to cooperation and conflicts among locals, as well as with outsiders, in the context of use of natural resources.

The last module of people's biodiversity registers deals with local suggestions on sustainable use and conservation priorities and the various measures needed to implement them. State policy, legal systems (those governing land, for example), harassment of women by officials, social norms about what is appropriate activity and behaviour for women-all play a role in excluding women.

Markets themselves, however, are hardly gender blind as their organization in most rural settings restricts access of many groups to contracts and information through shared systems of trust but also of exclusion. Deregulation may or may not open up markets to the excluded, but experience has shown that liberalization often intensifies exclusion. Policy attention to information, R&D results, extension, storage facilities, among other things, is essential to truly make markets effective and equitable allocative arenas. R&D thus needs to be situated in a real and existing social and economic system. It needs to look well beyond production technologies alone although the opportunities there are truly staggering as suggested by experience with agroforestry, horticultural tree crops, alley cropping and other hedgerow methods, living fences, and a wide range of water-harvesting and soil-contouring techniques.

Agricultural R&D financed and directed by the private sector, propelled by the profit motive, fuelled by market power, and constrained by an agenda set by interests outside the region, cannot deliver the goods. Their methodology is deficient, being based upon reductionist science, which breaks down the components of a complex world into discrete parts, analyses them separately, and then makes predictions about the world based on interpretations of these parts. Their motives are suspect, being driven by the capitalist imperative to grow and concentrate power, and their environmental and social sensibilities are primitive and laggard. The State must be deeply involved in agricultural R&D and in regional approaches to developing an alternative agriculture.

Providing employment guarantee schemes that function not only as social safety nets but also as work programs to provide rural infrastructure for environmental improvement. Employment- generation schemes must be integral to rural development and emphasize developing skills and infrastructure as well as providing wages in times of need. Effective agents of change cannot pick and choose their favourite items from the above list. They must take an integrated approach to agricultural transformation.

For example in Maharashtra has had a highly successful employment guarantee scheme since 1975. It has provided an average of 132 million work days on productive projects in soil and water conservation, small-scale irrigation, reforestation, and local roads. About half of employment guarantee scheme employment has gone to women. The employment guarantee scheme has been successful because (i) the right to employment is enshrined in state law, (ii) courts have upheld the right, (iii) funds are provided through earmarked taxes (mainly on professionals) and state government contributions, (iv) relative transparency governs the schemes operation, and (v) employment guarantee scheme job seekers are mobilized to demand their rights.

Sustainable agriculture conserves resources increases the ecological efficiency of farming to get more out of fewer resources while creating fewer negative side effects. It does so through greater resource productivity and through use of closed-loop systems of production that mimic natural processes. The creation of new local institutions that increase the market strength of small farmers, and the presence of State policies that level the playing field permit the powerless to compete and make sustainable agriculture economically feasible. Ultimately, sustainable agriculture will lead to a new kind of agriculture that eschews waste and external inputs entirely and is based on cooperation rather than competition with nature.

The following are some attributes of sustainable agriculture:

— Incorporation of natural processes into agricultural production, including nutrient cycling, nitrogen fixing, and pest-predator relationships.

— Reduction or elimination of use of off-farm, external, and nonrenewable inputs, starting with those most harmful to farm and community health and to the environment.

— Equitable access to productive resources and opportunities.

— Extensive use of local knowledge and practices, including innovations developed in isolation from modern science and technologies. Self-reliance among rural people.

— Use of complex cropping patterns, trees, shrubs, fish, and animals suited to local environmental constraints and opportunities.

— Sophisticated use of simple techniques of water harvesting, windbreaks, and soil and water conservation for long-term sustainability.

— Integrated farm management from a systems perspective that considers how to reduce waste, increase production, reduce risk, and improve livelihoods.

— Build the basis for urban agriculture as a contributor to food production, nutritional enrichment, livelihood development, aesthetic enhancement, vulnerability diminution, and political empowerment. These issues are further elaborated in the following sections on (i) land reform, (ii) rural financial services, (iii) common property resource management, (iv) irrigation management, and (v) coastal zone management.

In the last quarter of the 20th century, the reality of poverty and environmental degradation, however, can no longer be so blithely dismissed, and the subject of land and its role in sustainable development is once again on the policy agenda in Asia and the Pacific. A number of reasons to be concerned about land and land reform have to do with efficiency, equity, empowerment, and environmental sustainability. They have strong poverty, gender, social-exclusion, and quality-of-life dimensions.

Many of the issues are relevant to urban as well as rural areas. Simple solutions do not exist, and facile generalizations are nearly always wrong, except for one: land will continue to be one of the most urgent issues in rural Asia in the early 21st century. Because, first, the importance of land reform to economies having impressive growth is undeniable. Second, there are strong efficiency arguments for focusing on land use in agriculture. Economy of scale is rarely, if at all, possible in agricultural production as long as somewhat competitive markets for inputs and outputs exist, including rental markets for farm draft animals and machinery. Small farmers are at a disadvantage because of their greater transactions costs in imperfect markets and lack of access to financial and extension services.

The goal of government policy should therefore be to increase competition and access. However, the large farms need to supervise hired labour, which is difficult. A negative relationship between farm size and productivity exists in many countries and farming situations.

Efficiency and equity work hand in hand as long as small farmers are also ensured access to inputs and services. The history of land reform is littered with failures due to farmers' lack of access to financial services, appropriate extension services, and marketing opportunities.

In the Philippines, for example, farmers have often been forced to sell the land allotted to them back to large landowners because they could not pay for the necessary inputs. As poverty is concentrated among the landless and near-landless, land redistribution can have a significant impact on the incidence and severity of poverty. Farmers who own land not only have a productive resource, but also access to other resources that increase agricultural productivity. Lack of collateral prevents individuals from making investments that can be financed only with credit, including quality schooling, bullocks, wells, or perennial crops.

Gender equity is strongly influenced by landholding patterns. One related issue concerns intra-household effects of giving women more access to economic resources, as would be the case if they had control over land. Another issue relates specifically to land and how its control affects women. As far as the first issue is concerned, more economic resources in the hands of women-in their direct control and not mediated through the husband-translates into better nutrition, schooling, and health care for girls. It usually also translates into better household nutrition in general as women are more likely than men to convert economic resources into food.

Poor women normally spend all of their earnings on the household while men nearly always spend a large part of their earnings on themselves. Yet another important issue is what women do with land under their control. Given general and widespread discrimination against women, any woman-directed land reform has to be accompanied by reforms in complementary sectors such as agricultural extension, marketing, and financial services. Women may very well crop land in ways different from men's, which is likely to have a positive effect on the environment.

While men normally prefer cash crops if given a choice, women, whose primary interest is the family, may well opt for a diversified mix of cash crops, food crops, vegetables, fish, and small animals, the sum of which will improve household nutrition and reduce vulnerability to unexpected changes in weather or markets.

Nutritional welfare and ability to survive periods of stress, even famines, is very much affected by access to land. Access to land is considerably more important than redistribution of income equivalent to the output value of the land because land is a substitute for insurance and an efficient source of self-employment.

Communal tenure systems in many parts of the region periodically redistribute land precisely because the insurance benefit is so high. Women's right to land can have profound effects on women's rights in general. Land is a potent symbol of power, and studies in a number of countries have shown that where women have the right to land, they are less subject to violence from their husbands or other family members.

Men and women who have the right to land enjoy the respect of general society-whether represented by a local political leader, a government civil servant, a banker, or a nongovernment organization worker. The more evenly rights to land are distributed, the more even the playing field for economic development. One of the most important land reforms in the region is to identify forestlands that have been used for agriculture for many generations and to regularize their ownership.

Most forest-land statistics in Asia are a fiction, as large tracts of land have no forest cover at all. By refusing to allow the tillers secure tenure over these lands, however, there is an understandable lack of willingness to invest in sustainable agricultural improvements, such as terracing, small dams, or soil erosion control structures.

The land redistribution empowers the community, as it can only happen through collective organizing. The process of successful organizing, whether by groups of women, socially excluded castes, indigenous groups, or underprivileged classes, can bolster individuals' and groups' confidence in their abilities and worth. It can be the catalyst that may lead to freedom from poverty and social exclusion.

In the absence of social revolution, mode of land reform must adopt an experimental attitude and a pilot-project approach. In general, however, land reform may take the following forms:

— land redistribution;

— tenancy reform;

— abolition of intermediaries (which is still an issue in Pakistan, for example);

— land consolidation, which is difficult to implement and may not be as important as was once thought, given the implications of land fragmentation for the dispersal of risks;

— land titling, which will not, it must be emphasized, by itself help the poor and may, in fact, reinforce the privileges of the better-off;

— progressive taxation, which can help finance land redistribution, curb land speculation and the idle and unproductive use of land, and support decentralization and effective local government units;

— recognition and enforcement of gender rights, which are key elements of a successful land reform aiming for equity and efficiency;

— development of functioning land markets, which, however, by itself will not transfer land to more efficient producers or create greater equity; and

— market-based land redistribution. Any serious and effective land reform has to be State led and politically driven by demands from the poor themselves.

Financing is an obvious key, and the role of donors is clear and vital. Few small farmers can purchase land and finance their mortgages out of the profits of the first few years. It is also crucial to finance the development of new and sustainable agricultural technology, financial institutions, and marketing systems. One approach is to reallocate the massive government subsidies now going to the better-off farmers to a grant program for poor farmers or farmer groups so they can purchase land on the market in an open and transparent manner.

The beneficiaries, however, must be intimately and actively involved in the process, with the technical support of NGOs and/or government line agencies that can help them formulate farm plans and realistic farm investment proposals. Farmer groups would also be advocates for rights of their members and conduits for providing extension and marketing information. These process of land reform can be part of a process of agricultural transformation toward the new green revolution. It can be based on (i) community land-use planning, (ii) integrated model-farm planning, (iii) provision of a package of inputs and services, and (iv) a holistic approach to agriculture that aims for both equity and environmental sustainability.

Communal tenure systems do not need dismemberment but support in (i) increasing accountability and effectiveness of cooperation, (ii) responding to external political and market pressures, and (iii) devising sustainable technologies. Communal tenure systems can break down, and when they do, it is the most vulnerable and least powerful that lose. The challenges are (i) to build security for households into communal systems and (ii) to encourage accountability of political leaders to the needs of the community.

Essential inputs to agricultural transformation are credit, savings, insurance, financial advice, and training. The need for credit arises from the seasonal nature of agriculture. Inputs are needed at the beginning of the crop season, while payment is received at harvest. Credit is required to finance input purchases and longer-term investments in perennial crops, draft animals or machinery, storage facilities, livestock, irrigation facilities, and soil and water conservation works.

Sustainable agriculture, based as it is on mimicry of natural processes, requires a great deal of investment in conserving soil, water, and fertility. Mobilization of local resources, so essential to sustainable development, implies mobilization of local savings.

Reliance on the better-off and on resource extraction to fuel economic growth has had short-term success in terms of GDP growth. But the wreckage left behind as environmental destruction, social dislocation, cultural disempowerment, and poverty is now too foul to ignore. Only when the powerless cease to be so will the cycle be broken and poverty a memory of the past. Many studies show that it leads to increased incomes among credit recipients in the region.

For example, the first loan issued by an International Fund for Agricultural Development (IFAD) project in Indonesia raised household incomes by 31 percent. The second and third loans increased the cumulative gain to 72 percent. An early assessment of a Bangladesh Grameen Bank program showed real household income gains of around 35 percent. While the attribution of all of these gains to credit alone may be questioned, it is undeniable that credit has had a significant impact on people's lives. Credit also has an important role to play in smoothing consumption and mitigating the impact of shocks to the household economy. Thus, credit plays an important insurance function.

Credit also appears to alter the expenditure patterns of households-an effect that is especially significant when credit is advanced to women who tend to favour such expenditures as schooling and contraceptives. Total household consumption also rises more when credit is advanced to a woman than to a man. Microcredit programs rarely reach only the poorest, but they do reach poor households and many of the most vulnerable in society. It is these households that face the most difficulties coping with income shocks.

Credit has an important effect on the empowerment of women. An IFAD review of Nepal's Production Credit for Rural Women Project found that female beneficiaries felt less housebound, more confident in interacting with the outside world, and more independent because they had their own cash incomes. They also had more access to extension services. Credit can also change power relations within the household, which can have both positive and negative impacts on women.

When women cannot obtain loans or have to wait too long for a loan, for example, male impatience can lead to violence. It needs to be emphasized, however, that the positive impacts of credit arise because other inputs are available. The literature is full of bickering over the significance of this or that regression analysis showing this or that impact of credit on various outcomes. Access to markets, a certain level of skill development, and information about opportunities are essential as well. It is also highly likely that, especially in the case of microcredit to women and to the poor in general, group-mobilizing efforts themselves are a major contributor to positive outcomes.

Sustainability requires longer-term investments in soil and water conservation and soil fertility management. It also requires a diversified cropping structure that includes perennials. Credit is necessary not only for the investments themselves, but also for consumption needs during the transition period.

Grameen Bank in Bangladesh, for example, has a savings program but does not aggressively reinvest the savings in rural development. When the State was the focus of development efforts, agricultural development banks were all the rage. They provided subsidized credit for farmers needing access to green revolution input packages. They made many attempts to target credit to the poor, which seemed doomed to failure until the invention of group-lending microcredit institutions. Rural credit open to individuals at subsidized rates was monopolized by the better-off, no matter what the targeting mechanism. When credit was not sufficiently subsidized, borrowers merely defaulted and faced no penalties for doing so because they belonged to the elite.

Unfortunately, as microcredit has shown its success in actually reaching the poor, the better-off invoke sustainability criteria to show that it is nonviable and in conflict with market principles. The poor lose out once more, first to the State and now to the market.

Grameen Bank is also good at what it does, although it may need to make changes to improve its investment of group savings, modestly increase interest rates, and perhaps further improve efficiency. It seems wrong-headed to insist that it be financially sustainable given its excellent record in actually reaching the poor. But it does need to be emphasized that subsidies ought to be spent on institutional strengthening, expansion of operations, and social mobilization rather than on creating a dependence on cheap credit while crowding out private financial service initiatives.

If Grameen Bank were more selective in its targeting to better its financial performance, then it would not reach the poorest of the poor. If, however, Grameen Bank were to be more concerned with reaching the poorest, then it would inevitably be criticized for its financial unsustainability. The point is that Grameen Bank does reach the poor. Almost all of its clients are poor. That in and of itself is a remarkable accomplishment. Emphasis on financial sustainability and credit performance measured by high repayment rates has incentive effects on fieldworkers, which work to the disadvantage of the poorest households and even more profoundly to the detriment of women. As loans become larger with mature borrowers, men increase their control over loans although they may still be in the names of women.

For women to retain and increase their control over their own lives requires not just access to microcredit, however important that may be, but also improvements in women's productivity, mobility, access to markets, literacy, social status, and control of household decisions. These all take time and require (i) the building of women's institutions, (ii) the provision of technology and training support for women's investments, and (iii) women's consciousness raising.

In many instances, it may be that households want a secure savings scheme for consumption smoothing rather than credit per se. In other instances, the group support and input can make a big difference in outcomes to credit access. To the extent that financial management is one of the services offered, then it can often become as important as credit itself in helping move households out of poverty.

Microfinance is not well suited to many types of agricultural investments and is, in fact, often used for rural nonfarm activities. But it can be successfully employed in intensifying low-input agriculture such as organic vegetable farming and marketing or low-input integrated livestock-crop-fish-fodder systems.

The reason is that credit alone, especially credit that is mostly nonagricultural in purpose, is marginal to the major earning opportunities operating in the rural economy: (i) agriculture; (ii) dynamic rural small industry; and (iii) trading in food grains, agricultural inputs, and commodities produced and consumed in the rural economy. Agricultural credit remains the domain of the traditional moneylender.

Rural productive opportunities remain tightly tied to landownership and control of productive assets. The landlord, the moneylender, and the merchant have the land, capital, skills, and access to State resources that are the truly determinant factors in rural poverty formation and maintenance.

The growth elasticity of poverty varies widely across countries. Country comparisons imply that it is indispensable to pay close attention to the ability of poor and socially excluded households and individuals to participate positively in the growth process. "Participation" depends on access to resources, education and other social services, technology, and markets. The social relations and the resultant institutions that govern such access are of vital importance.

Common property resources have signal importance for the poor and women. Certain classes of common property resources are also among the most degraded lands in Asia. Thus, they highlight common causes of poverty and environmental deterioration. A useful simplification is to consider four broad types of landownership and use: (i) private; (ii) open access (where well-defined property rights are absent); (iii) State; and (iv) communal. All may involve some degree of common property rights. Even private property has historically and to this day been subject to rights of gleaning, with poor persons having "rights" to glean from private fields and orchards. Common property resources (forests or woodlots, pasturelands, coastal fisheries, inland waters, groundwater) are being degraded and closed to common access through privatization or takeover by the State. Both processes worsen access of the poor and of women.

There are many examples of traditional, local methods of management of common property resources. Institutional mechanisms have not only developed in many different circumstances, but also altered over time to adapt to changing conditions. However, they are obviously not impervious to shock if external conditions change drastically.

Local resource control systems, social sanctions to protect the community's stake, and enforcement mechanisms worked effectively to protect the environment. The State saw the revenue and control possibilities in common property resources and often nationalized communal and private lands, sometimes for their own "protection." Market penetration, breakdown of cooperative values, land reform, and centralized uniform thinking led to privatization of communal and open-access lands. The process was led by and benefited the better-off and men.

In the end of the 20th century saw a proliferation of initiatives to restore some measure of environmental integrity to degraded forest and pasturelands and to improve management of irrigation systems. Similar efforts in coastal fishing communities are more recent, but high on the policy agenda. User groups have been advocated by development professionals, and many examples now exist of successful user-group formation and management of natural resources.

Some user groups have built upon preexisting traditional community institutions, but many have been constructed from scratch. The good news is that user-group formation has achieved some impressive successes in halting environmental degradation and in managing water resources more efficiently.

Unfortunately, their successes have often (but not always) come at the cost of women and poor households, who are systematically excluded from access to common property resources, which have been vital to the livelihoods of large numbers of the poor.

In India alone, common property resources add approximately $5 billion a year to the incomes of poor rural households. An average of 12 percent of poor rural households' income is from common property resources. In some areas, poor households derive considerably more of their total income from them. Fuel and fodder are the most important common property resources accessed by poor households, and women and children are responsible for their collection. All studies on India agree that common property resources are being degraded, reduced in area, and systematically closed to the poor.

In Nepal and India, many examples abound of green hillsides regenerating forests where a few years ago there was only brown scrub. "Wastelands" have been transformed across India. Nepal has not been completely deforested as some of the more hysterical international observers predicted a quarter century ago. A number of hill regions now have more forest cover and a higher quality of forest than they did 20 or even 40 years back.

In many parts of Asia and the Pacific, incomes have increased, out-migration has eased, water has become more plentiful in streams and wells, and biodiversity has increased. They are remarkable accomplishments, but mixed blessings at best for women and poor households. Before discussing social forestry on common lands, it would be useful to point out that some private lands have also seen significant increases in tree cover and total biomass production. In Nepal, for example, a great deal of the increase in tree cover has occurred on private land. While more beneficial to larger landowners, it may be that with the increase in stall feeding and a reduction in pressure on common lands, the distributional and gender effects of private land tree growing have not been all negative.

In Java, once-barren hillsides have been transformed into fruit orchards to supply the protected and growing Jakarta market, which has led to a reduction in soil loss from erosion. Four other effects, however, mitigate the income gain and erosion reduction. First, the "apple lords" are all men, and all formal contractual agreements are between men. Women may do much of the work and often are managers, but men retain control over cash incomes. Second, the fruit boom has accelerated social and economic differentiation, with the power of the "apple lords" increasing dramatically. Third, agro-forestry as practiced in the Java apple and orange orchards is not only favourable to large growers and to men, but is also heavily reliant on external inputs. It is a far cry from the traditional nutritionally efficient Javanese home garden that recycled nutrients, provided a balanced diet, was a form of insurance against economic and weather crises, and was controlled by women. Finally, monocrop fruit orchards dependent on protected Indonesian urban markets are highly vulnerable to pest problems, liberalization of fruit import restrictions, and financial crises.

Community-based forest management is the latest rage in many countries of the region. Although the details differ from country to country, the rhetoric calls for (i) ending the State monopoly of control over significant tracts of forestland, (ii) recognizing how community needs and knowledge about forest management can complement scientific forestry management, and (iii) relinquishing control over forest management to community- based user groups. The rhetoric is full of "participation," "environmental sustainability," "cooperation," and even "equity" and "gender sensitivity."

Studies show that joint forest management in India and community-based forest management in the Philippines are in reality far closer to the old style of scientific State-managed forestry than to an ideal co-management model. In both countries, the State sets up its own explicitly exclusionary village forest protection committee or people's organization independently of and sometimes in competition with already-existing community institutions. Whereas agency targets used to be in terms of trees planted, now they are people's organizations established.

People's participation is implicitly assumed to be involved, but the details are left to NGOs and other contracted organizations. Despite much talk of ownership of stakeholders, in fact ownership and ultimate control rest in the hands of the State. In general, these management efforts are restricted to already-degraded pasture and wastelands and designed for peasant subsistence use. Program support tends to be narrowly focused, reflecting line agency priorities, in contrast to farmers' traditional integrated strategies employing annual-perennial mixes, livestock raising, and balances between extensive and intensive land uses.

In the Philippines, for example, social forestry or community-based forest management programs continue to be strongly oriented toward efficiency and productivity, and without social and broad environmental goals. The State controls the programs. A community is understood as a geographically defined and homogeneous group of like-minded people who have to be organized by State-directed NGOs into people's organizations.

In the Philippines, India, Nepal, Thailand, and elsewhere, where social forestry schemes involving State-community co-management of forests have been successful in increasing marketable forest products, the lure of quick profit almost always leads to the control of revenues by the male nonpoor. Rarely do women or poor households reap cash benefits from the sale of products from the managed forest.

In India, joint forest management maintains the principle that the State is the sole proprietor of forestlands, and does not recognize any community-based rights. It makes no provision for grant or lease of land to communities. Joint forest management agreements are revocable by the Forest Department at any time for any reason.

Local participation in planning is a part of the rhetoric of joint forest management, but State resolutions on the subject do not require public participation. As long as the areas in question are badly degraded and devoid of marketable timber, these issues may not matter so much. But when they become productive forests, competition for their control will intensify and the State will be able to assert its control over them. Women lose the most from these changes, especially in South Asia. Their fuel and fodder collection time is increased, and children must be kept out of school to help out. Of course, it is not inevitable, and the theoretical models call for community-managed forests to have management plans that reflect the diversity of needs of the entire community.

Many joint forest management rules provide for women's participation in committees. Forest-users' groups often require the committees to have women members. The committees talk a great deal about fodder species, fuel collection policies, and minor forest products. In truth, however, user groups usually consolidate the control of males and the better-off households. Joint forest management committees in India are overwhelmingly male. In West Bengal, women constitute 3 percent of over 8,000 members of joint forest management committees.

In Tamil Nadu, 7 percent of over 22,000 members are women. Even when women attend meetings, they rarely speak, and when they do, they feel they are not heard. Rules that prohibit their collecting fuel and fodder increase their workload. When management sells the forest products formerly collected by poor households, it provides cash to males and more work and hardship to the excluded. Cash in the hand of a male household head is less likely to be spent on household necessities than on his personal consumption. The outcome is not inevitable, however, and therein may lie some important lessons for sustainable development.

In essence, the bargaining power of the socially excluded, the poor, women, and indigenous groups must be enhanced relative to that of the male elite. What is important is that the disadvantaged groups have (i) better economic and legal status, (ii) improved external support, and (iii) an enlightened ideology about appropriate social norms, rights, roles, and abilities.

Women are major users of common property resources. With the help of children, they collect fuel, fodder, and water, as well as a great deal of food, medicine, and building material. Unfortunately, their current role as collectors is translated directly into their new role as protectors by those who believe that if the environment is protected and productivity is enhanced, then women must benefit. Unfortunately, they ignore the fundamental rights of women to use common property resources.

They also ignore the already-overwhelming work burden placed on the backs of women and children by poverty, inequitable resource access, technological inefficiencies, lack of alternative fuel (kerosene, for example) and fodder sources, and lack of convenient and safe drinking water supplies. Fortunately, some initiatives in common property resource management do include women's gaining a greater share in resources.

Government financial shortages and donors' philosophical commitment to reducing the role of the State in developing countries have led to the creation of many programs devolving irrigation management responsibility to water-users' associations. It is generally asserted that the associations will increase water-user participation in decision making and that cost recovery, water-use efficiency, and agricultural productivity will increase as government subsidies to irrigation decline. A vast literature has sprung up on the subject of water-users' associations, mostly written by advocates of devolution. Strikingly few detailed, comprehensive, and quantitative studies are part of the literature, however, as authors rush to publish their breathless anecdotes of farmer management abilities.

Impacts on the environment, women, equity (inter-farm water allocations between head and tail of an irrigation system and access to water of the smallest farmers), farmer costs, water-use efficiency, and changes in cropping patterns and crop productivity are infrequently reported, and when they are, the accounts are usually qualitative. The push for devolution of irrigation services in the region has rarely arisen from either farmers themselves or from irrigation departments, but has been driven by donor pressure and government budgetary shortfalls. The result has been hastily designed and implemented devolution programs often reflecting neither field realities nor a considered appreciation of the requirements of a successful devolution. Studies show that results have been both positive and negative.

In Asia, impacts on women are found to be generally neutral to negative. Impacts on equity head-versus-tail are often not reported, and when they are, are done so anecdotally. Costs to government have decreased, although the only fairly comprehensive studies of this impact have been for the Philippines National Irrigation Administration where financial savings to Government were substantial in the early years. Head-versus- tail water deliveries, for example, have been generally positive, but negative in some Philippine schemes.

Sri Lanka, Indonesia, and Nepal also saved money on turnovers of maintenance responsibilities to water-users' associations. Not surprisingly, costs to farmers have increased, often substantially, and, in some cases, the facilities have been abandoned as financially unenviable. On the other hand, farmers have often cut operation and main-tenance costs substantially and sometimes made pump use more effective, saving on pumping costs.

The coastal zones of Asia and the Pacific are under pressure from unsustainable resource exploitation; municipal, marine, and industrial pollution; and increasing risks and impacts of natural hazards such as cyclones, typhoons, tsunami, and flooding. Traditional means of coastal zone management based on community management and sophisticated rules and sanctions have, for the most part, been rendered ineffective by powerful economic, institutional, technological, and political changes. Although consumption of fish and other marine and coastal products is declining among low-income households in many Asia-Pacific developing countries, worldwide market demand for fish, prawns, and other marine resources has surged.

Seafood has become a health and status food in industrialized counties. One consequence of this has been the dramatic depletion of fish stocks in the inland, coastal, and marine waters of the industrialized nations. Fishing fleets armed with new technologies for intensive fish harvest and driven by increasing market demand have thus combined with increases in the number of fishing vessels in developing countries to lead to overfishing of most species in most coastal waters. Industrial and municipal pollution, destruction of mangroves and corals, tourism and urban development, mining, erosion from deforested and cropped lands, and high demand for shells, coral, and aquarium fish all combine to exacerbate pressure on coastal resources caused by international market and technology changes.

In the Philippines, for example, total fish production has been decreasing by about 2.5 percent per year since 1987. Fishers today spend about 10 times more resources to collect the same amount of fish that similar operations did 50 years ago. Most municipal fishers operate at a loss and keep at it only because they value their labour at near zero and have no other skills to compete in the employment market, which is particularly tight in poor coastal communities. Municipal fisheries have been the hardest hit by production declines, and because 68 percent of workers in the industry are municipal fishers and rank among the poorest households in the country, this decline severely impacts both livelihoods and nutrition.

Municipal fisheries are normally the most productive fisheries because of the presence of reef, mangrove, and seagrass ecosystems, but they are also especially vulnerable to pollution, land-based erosion, and mangrove and seagrass destruction. Municipal fishers face deteriorating ecosystems and the more efficient technologies of commercial trawl fishers in bays once reserved for them. For example, trawlers have fished out San Miguel Bay, Philippines, beyond the point of natural recovery.

Coastal waters that were once a community asset have become open-access resources to anyone who can afford craft and gear. This is true from Kerala, India, to the South Pacific. Municipal fishers often resort to desperate measures, using dynamite, fine mesh nets, and sodium cyanide, all of which merely accelerate the decline of the fishing grounds. A related issue is hazard management. As unplanned and uncontrolled developments occur in coastal areas vulnerable to natural disasters, and as the role of the State in regulating development diminishes, institutional forms of hazard management and response become moribund and new institutional mechanisms are slow to meet the need.

In Viet Nam, for example, locally organized collective action for coastal defence and water management has been undermined by the dismantling of communal institutions such as agricultural cooperatives and the replacement of central and provincial planning and collective action with individual responsibility. As activities such as dike maintenance become increasingly monetized and professionalized, collective responsibility and local participation in a common task give way to uneven results.

Some districts tax and spend the proceeds on dike maintenance; the poorer ones may not do so. Local autonomy and privatization have contributed to considerable loss of mangroves in coastal Viet Nam, further weakening coastal defences. Thus, in a very short period of time, a highly effective coastal management system has withered under economic reforms and is yet to be replaced by a suitable alternative. There are, however, encouraging signs of a turnaround where communities have been able to mobilize fishing households to stand up for their rights against encroachers and where these communities have been supported by a broader coalition of forces whose interests are served by stronger and more effective management of the coastal zone.

Small-scale illegal fishing has declined although sustainable fishing rates may still be exceeded. Monitoring is ongoing. Solid waste is no longer dumped directly into the bay by the municipal governments and there is a notable increased awareness on the part of local government units of the importance of coastal pollution to the marine environment. Much remains to be done but progress is encouraging. However, the bay communities have yet to deal with deforestation and its attendant erosion and sedimentation, and while awareness of the linkages with fisheries is increasing, the necessary political coalition to bring about changes in land use has yet to be built.

Bibliography

Apu, N.A., and Middendorp, H.A.J. (1998) *Establishing fishers groups for self-management of enhanced fisheries in semi-closed water bodies in western Bangladesh*. The experience of the Oxbow Lakes small-scale fisherman project, in T. Petr (ed.) 'Inland fishery enhancements', paper 374, pp.393–406. Rome: FAO.

Carney, D. (ed). (1998) *Sustainable rural livelihoods: What contribution can we make?* London: DFID.

Chambers, R. (1987) *Sustainable livelihoods, environment and development: Putting poor rural people first*. IDS Discussion Paper. No. 240. Brighton: IDS.

Edwards, P., and Demaine, H. (1997) *Rural aquaculture: Overview and framework for country reviews*. Bangkok: RAP/FAO.

Edwards, P., and Demaine, H. (1998) *Completing the problem-solving cycle in aquaculture: The AIT Aqua Outreach*. *Aquaculture Asia*, 3(3): 10–17.

Edwards, P., Demaine, H., Innes-Taylor, N. and Turonguang, D. (1996) *Sustainable aquaculture for small-scale farmers: Need for a balanced model*. *Outlook on Agriculture* 25(1): 19–26.

Eknath, A.E., and Acosra, B.O. (1998) *Genetic improvement of farmed tilapias (GIFT) project*. Final Report. Manila: ICLARM.

FAO (1999) *Asia-Pacific Fishery Commission*. FAO Fisheries Report 610 of the Ad Hoc Working Group of Experts in Rural Aquaculture, Bangkok, pp.22. Rome: FAO.

Little, D.C., Innes-Taylor, N.L. Turongruang, D. and Komolmarl, S. (1991) *Large seed for small-scale aquaculture*. *Aquabyte*, 4(2): 2–3.

Moura Costa, P. and Stuart, M. (1998) *Forestry-based Greenhouse Gas Mitigation: a short history of market evolution*, Commonwealth Forestry Review Vol. 77 (3): 191–201.

Moura Costa, P. H. and Wilson, C. (1999) *An equivalence factor between CO_2 avoided emissions and sequestration – description and applications in forestry*, Mitigation and Adaptation Strategies for Global Change (in press).

Pearce D., Day B., Newcombe J., Brunello T. and Bello T. (1998) *The Clean Development Mechanism: Benefits of the CDM for Developing Countries*, Final Draft produced by CSERGE under contract to DFID. The Centre for Social and Economic Research on the Global Environment, University College London.

Scoones, I. (1998) *Sustainable rural livelihoods: A framework for analysis*. Working Paper 72. Brighton: IDS. DFID (1997) The UK White Paper on International Development - and beyond. London: DFID

Stuart, M. and Moura Costa, P. (1998) *Greenhouse Gas Mitigation: A Review of International Policies and Initiatives*, Policy that works for forests and people series, Discussion Paper No. 8. London: IIED.

Van der Mheen, H. (1998) *Achievements of smallholder aquaculture development in southern Africa*. Experiences from the ALCOM Aquaculture Programme. Harare: FAO.

Index